GENETIC ENGINEERING, FOOD, AND OUR ENVIRONMENT

To Bradley
great b... ...
rock on
Luke Anderson

W9-BFG-028

Luke Anderson lives in South Devon,
England. He writes, campaigns and
speaks internationally on issues
related to genetic engineering.

GENETIC
ENGINEERING, FOOD, AND OUR ENVIRONMENT

Luke Anderson

CHELSEA GREEN PUBLISHING COMPANY
WHITE RIVER JUNCTION, VERMONT

First published in the United Kingdom in 1999
by Green Books Ltd

First published in the USA in 1999 by
Chelsea Green Publishing Company,
P.O. Box 428, Gates-Briggs Building
White River Junction, Vermont 05001
(800) 639-4099
www.chelseagreen.com

Updated reprint March 2000

Cover design by Rick Lawrence

Printed by J.W. Arrowsmith Ltd
Bristol, UK

Text printed on Cyclus Offset 100%
recycled paper, manufactured from
waste paper. 100% of residuals
from production are reused.
Non-chlorine bleached.

U.S. Library of Congress Cataloging-in-Publication
data available on request.

ISBN 1 890132 55 1

CONTENTS

"The idea that we live in something called the environment is utterly preposterous . . . environment means that which surrounds or encircles us; it means a world separate from ourselves, outside us. The real state of things, of course, is far more complex and intimate and interesting than that. The world that environs us, that is around us, is also within us. We are also made of it; we eat, drink and breathe it; it is bone of our bone and flesh of our flesh."

Wendell Berry

Introduction

Much of the food now eaten in industrialised countries, especially in the United States, contains genetically engineered ingredients. Although there are powerful commercial interests behind the introduction of these foods, and support for the genetic engineering industry from many national governments, there is growing opposition from the public.

Throughout 1999, a string of new scientific papers, biotech company reports, political scandals and public protests fuelled extensive media coverage of the introduction of genetically engineered crops in food and agriculture. In May 1999, for example, a study published in the scientific journal *Nature* revealed that caterpillars of the Monarch butterfly suffered from retarded growth and increased mortality after eating leaves dusted with pollen from a widely grown variety of genetically engineered corn.[1]

In July, a report written by Deutsche Bank, one of the world's leading investment firms, stated that GE products were fast becoming an economic liability. Food processors were paying premiums of up to $1 per bushel for "non-GE" commodities, and GE ingredients were being withdrawn from major supermarkets, food producers, beverage companies, animal feed suppliers and restaurants across Europe, and also in Japan, Australia, New Zealand, Korea, Thailand, the US, Canada, South Africa, Mexico, Brazil and Hong Kong.[2] And in the US, where farmers lost nearly $2 billion in soya and maize exports to Europe between 1996 and 1998,[3] over 30 farm groups, including the National Family Farm

Coalition and the American Corn Growers Association, were now advising farmers in the US not to grow GE crops.[4]

By the time governments came together in Seattle for the fated World Trade Organisation meeting in November 1999, genetic engineering was one of the issues at the top of the international political agenda. Inside the meeting, governments from around the world expressed their concerns that the GE exporters might force them to accept genetically engineered products against their will. Outside on the streets of Seattle, where thousands of protestors were braving pepper spray, tear gas and rubber bullets, banners saying 'No to Genetic Engineering' hung alongside others criticising the impact of 'free trade' on the environment and on the poorest countries of the world. Two months later, despite the powerful opposition of GE-exporting countries such as Canada and the US, 130 countries signed an agreement called the Biosafety Protocol, which gives them the right, based on the application of the 'precautionary principle', to refuse GE imports.

This book aims to explain why genetic engineering has now become such a critically important issue, by providing an introduction to the social, environmental and health implications arising from the commercial use of the technology in food and farming. I do not discuss human genetic engineering at any length, nor the use of GE in biological weapons. For further information about these matters, consult the 'Resources' section at the back of the book.

Luke Anderson
February 2000

N.B. In the text, the terms 'genetically engineered', 'GE' and 'transgenic' are used synonymously. British usage (e.g. for corn/maize) and spellings have been retained.

Erratum

In my enthusiasm to include the results from the January 2000 meeting of the Biosafety Protocol, I made a mistake in the introduction just before this edition went to press. On page 8, "130 countries signed an agreement called the Biosafety Protocol" should instead read "130 countries reached an agreement called the Biosafety Protocol". Although delegates from 128 countries did agree to the Biosafety Protocol at the meeting in January 2000, the signing of the agreement began in May in Nairobi. At the time of writing this erratum, 67 countries have so far signed the protocol.

Luke Anderson, May 2000

Chapter One

What is Genetic Engineering?

What is a gene?

Every plant and animal is made of cells, each of which has a nucleus. Inside every nucleus there are strings of DNA, and these strings of DNA are organised into structures called chromosomes.

Each cell normally holds a double set of chromosomes, one of which is inherited from the mother and one from the father. One set of chromosomes from each parent combines when the sperm fertilises the egg (in the case of animals) or pollen fertilises the ovum (in the case of plants). The cell formed after fertilisation divides into two identical copies, each of which inherits this unique new combination of chromosomes. These embryonic cells then continue to divide again and again. The inherited genetic material, carried in the chromosomes, is therefore identical in each new cell.

DNA is often described as a blueprint containing all the essential information needed for the structure and function of an organism; genes are described as the individual messages that make up this blueprint, each gene coding for a particular characteristic. Although this concept can be helpful as a tool for understanding, it runs the risk of reducing the organism to a machine, and viewing physiology as little different from a series of industrial processes.[1] In reality, genes are very difficult to define and can only be understood within their context—a living organism.

Genes are sequences of DNA; they operate in complex networks that are regulated to enable processes to happen in the right place and at the right time.[2] No gene works in isolation. "Genes are arranged along the DNA in groups or 'families'", explains Dr Michael Antoniou, a molecular biologist who specialises in the clinical applications of genetic engineering. "The function of a given gene in a group is dependent on all the other genes that are present within the same family. Furthermore, the genetic activity in one family of genes can affect the function of genes in other groups of genes. It is also clear that genes and the proteins that they give rise to, have co-evolved together to form an extremely intricate, interconnected network of finely balanced functions, the complexities of which we are only just beginning to understand and appreciate."[3]

This intricate network also responds to external influences. According to Barbara McClintock, who won the Nobel Prize in 1983 for her pioneering work in the field of genetics, the functioning of genes is "totally dependent on the environment in which they find themselves".[4]

What is genetic engineering?

It is said that genetic engineering is simply "the latest in a 'seamless' continuum of biotechnologies practised by human beings since the dawn of civilization, from bread and wine-making, to selective breeding."[5] Although it is true that the food crops we are eating today bear little resemblance to the wild plants from which they originated, there are clear differences between genetic engineering and traditional breeding.

In traditional forms of breeding, variety has been achieved through selection from the multitude of genetic traits that already exist within a species' gene pool. In nature, genetic diversity is created within certain limits. A rose can be crossed with a different kind of rose, but a rose will never cross with a potato. Even when species that may seem to be closely related do succeed in breeding, the offspring are usually infertile—a horse, for example, can mate with an ass, but the offspring (a mule) is sterile.

Genetic engineering, on the other hand, usually involves taking genes from one species and inserting them into another in an attempt to transfer a desired trait or character. This could mean, for example, selecting a gene which leads to the production of a chemical with antifreeze properties from an arctic fish (such as the flounder), and splicing it into a potato or strawberry to make it frost-resistant.[6] It is now possible for plants to be engineered with genes taken from bacteria, viruses, insects, animals or even humans.

> "Researchers in the field of molecular biology are arguing that there is nothing particularly sacred about the concept of a species . . . they see no ethical problem whatsoever in transferring one, five or even a hundred genes from one species into the hereditary blueprint of another species. For they truly believe that they are only transferring chemicals coded in the genes and not anything unique to a specific animal. By this kind of reasoning, all of life becomes desacralised. All of life becomes reduced to a chemical level and becomes available for manipulation."—Jeremy Rifkin [7]

It is often argued that there is no scientific foundation to concerns about these cross-species gene transfers, because there is evidence that bacteria and viruses have often transferred genes between species in the course of evolution. According to this line of thinking, genetic engineering just speeds up what is essentially a natural process. Rates of change, however, are highly significant: if this argument were applied to radioactivity, which is present at low levels all around us, one could similarly downplay the risk of the processes used in nuclear reactors, on the basis that that they simply speed up the rate of radioactive decay, a naturally occurring phenomenon.

The mechanics of plant genetic engineering

There are a number of techniques in the genetic engineer's toolkit. A biochemical process is used to cut the strings of DNA in different places and select the required genes. These genes are usually then inserted into circular pieces of DNA (plasmids) found in bacteria. Because the bacteria reproduce rapidly, within a short time thousands of identical copies (clones) can be made of the 'new' gene. Two principal methods can then be used to insert a 'new' gene into the DNA of a plant that is to be engineered.

1. A 'ferry' is made with a piece of genetic material taken from a virus or a bacterium. This is used to infect the plant and in doing so smuggle the 'new' gene into the plant's own DNA. A bacterium called *Agrobacterium tumefaciens* (which causes gall formation in plants) is commonly used for this purpose.

or

2. The genes are coated onto large numbers of tiny pellets made of gold or tungsten, which are fired with a special gun into a layer of cells taken from the recipient plant. Some of these pellets may pass through the nucleus of a

cell and deposit their package of genes, which in certain cases may be integrated into the cell's own DNA.

Because the techniques used to transfer genes have an extremely low success rate, the scientists need to be able to find out which of the cells have taken up the new DNA. So, before the gene is transferred, a 'marker gene' is attached, which codes for resistance to an antibiotic. Plant cells which have been engineered are then grown in a medium containing this antibiotic, and the only ones able to survive are those which have taken up the the 'new' genes with the antibiotic-resistant marker attached. These cells are then cultured and grown into mature plants.

A piece of DNA taken from a virus or bacterium (called a 'promoter') is also inserted along with the 'new' gene in order to 'switch it on' in its new host. Promoters, which often force genes to express their traits at very high levels, also have the potential to influence neighbouring genes.[8] The promoter may, for example, stimulate a plant to produce higher levels of a substance which is harmless at low levels but which becomes toxic when present in higher concentrations.

As it is not possible to insert a new gene with any accuracy, the gene transfer may also disrupt the tightly controlled network of DNA in an organism. Current understanding of the way in which genes are regulated is extremely limited, and any change to the DNA of an organism at any point may well have knock-on effects that are impossible to predict or control. The new gene could, for example, alter chemical reactions within the cell or disturb cell functions. This could lead to instability, the creation of new toxins or allergens, and changes in nutritional value.[9]

- A gene coding for red pigment was taken from a maize

plant (corn) and transferred into petunia flowers. Apart from turning red, the flowers also had more leaves and shoots, a higher resistance to fungi and lowered fertility.[10]

- In a trial of genetically engineered insect-resistant maize, there was an unexpected yield reduction of 27% and significantly lower levels of copper in the leaves, stalks, and grain compared with the control plants.[11]

- A yeast was genetically engineered for increased fermentation purposes. A toxic metabolite called methylglyoxal was produced in concentrations 30 times higher than in the non-genetically engineered strains.[12]

- In trials used to assess the safety of herbicide-resistant soybeans made by Monsanto, 36 cows were divided into different groups; for four weeks some were fed transgenic soybeans, and some fed with ordinary ones.[13] When the data from the trials were examined, it was found that the cows that were fed the normal soybeans produced 1.19 kg of milk fat a day, whereas those fed with genetically engineered soybeans produced 1.29 kg—an increase of over 8%.[14] This shows that a genetic change which was only intended to make a soybean resistant to a herbicide, had side effects—which have not been explained. No further tests were conducted to explore these changes, and the genetically engineered soybeans have been passed by the regulatory authorities as safe for consumption.

- In 1997, scientists at the University of Oxford who were investigating the metabolism of potatoes, unexpectedly found out how to use genetic engineering to increase their starch content. The scientists were working on what they believed to be a quite different aspect of potato metabolism, when they discovered that suppressing the activity of an enzyme dramatically affected the levels of starch produced within the potatoes. "We were as surprised as anyone," said Professor Chris Leaver, Head of the Department of Plant Sciences at the university. "Nothing in our current understanding of the metabolic pathways of plants would have suggested that our enzyme would have such a profound influence on starch production."[15]

"Up to now, living organisms have evolved very slowly, and new forms have had plenty of time to settle in. Now whole proteins will be transposed overnight into wholly new associations, with consequences no one can foretell, either for the host organism, or their neighbors. It is all too big and is happening too fast. So this, the central problem, remains almost unconsidered. It presents probably the largest ethical problem that science has ever had to face. Our morality up to now has been to go ahead without restriction to learn all that we can about nature. Restructuring nature was not part of the bargain. For going ahead in this direction may be not only unwise, but dangerous."—Dr. George Wald, Professor Emeritus in Biology from Harvard University and Nobel laureate in medicine [16]

The notion of 'substantial equivalence'

Many people became aware of GE food for the first time in 1996, when soybeans grown in the US were genetically engineered by Monsanto to be resistant to their best-selling herbicide Round-up. Over 40% of the US soybean harvest is exported, and when the first consignment of GE soya arrived in Europe, it was already mixed in with the conventional harvest. The American Soybean Association rejected calls to segregate the GE soya on the basis that it was 'substantially equivalent' to ordinary soya.[17]

The theory of 'substantial equivalence' has been at the root of the international safety assessment and testing of GE food. According to this principle, selected chemical characteristics are compared between a GE product and any variety within the same species. If the two are grossly similar, and if it is shown that the genetic engineering has not inadvertently led to the production of known toxins and allergens, the GE product does not need to be rigorously tested on the assumption that it is no more dangerous than the non-GE equivalent. None of the products that have been approved so far has undergone long-term safety tests.

> "Following the principles for the application of substantial equivalence, there should be no further safety or nutritional concerns of any significance."—Monsanto Application to the UK Advisory Committee on Novel Foods and Processes for Review of the Safety of Glyphosate Tolerant Soybeans, 1994

The use of 'substantial equivalence' as a basis for risk assessment is seriously flawed, and cannot be depended on as a criterion for food safety. It focuses on risks that can be anticipated on the basis of known characteristics, but ignores unintended effects that may arise.[18] Genetically engineered food may, for example, contain unexpected new molecules that could be toxic or cause allergic reactions. A product could not only be 'substantially equivalent', but even be identical with its traditionally produced counterpart in all respects bar the presence of a single harmful compound.

L-tryptophan

In 1989, a new brand of a widely used food supplement called L-tryptophan was put on the market in the United States. A company called Showa Denko had been experimenting with a new process using genetically engineered bacteria. Through the use of genetic engineering techniques, scientists at Showa Denko were able to cause the bacteria to increase the amounts of L-tryptophan which they produced.

After the release of this product, an estimated 5000 people suffered from an outbreak of a disease called Eosinophilia Myalgia Syndrome (EMS). Government authorities initially reported that 37 people died and 1500 were left with permanent disabilities,[19] including paralysis, chronic neurological problems, painful swelling and cracking of the skin, heart problems, extreme sensitivity to light and autoimmune disorders.[20] However, the authorities stopped monitoring deaths from the disease shortly after the outbreak; many more people with EMS are thought to have died since then.[21]

This batch of L-tryptophan had not been labelled any differently to non-GE batches, and it took several months before the US authorities realised that it was the source of the outbreak. There was a nationwide recall of the L-tryptophan, and it was found to contain minute traces of highly toxic compounds which appeared to have been novel byproducts of the production process.[22]

The exact cause of EMS remains controversial. Scientists are unable to reach any definitive conclusions because Showa Denko immediately destroyed all batches of the genetically engineered bacteria.[23] It is also known that Showa Denko was cutting corners in its purification processes, reducing the levels of carbon powder used to purify the L-tryptophan. However the company maintained that it was not unusual for the levels to be so low, and said that they had cut corners before without any apparent ill effect. If failures in the purification process were at the root of the problem, we could expect to have seen contaminants of this kind in non-GE batches of L-tryptophan. Although some contaminants have been discovered, albeit at much lower concentrations, one of the toxins (Peak E—also called Peak 97 and EBT) has never been found in brands produced with non-GE bacteria.[24]

Official bodies around the world have been quick to discount suggestions that genetic engineering could be to blame for the tryptophan incident. Regarded as 'substantially equivalent' to the L-tryptophan produced without the use of genetic engineering, this batch was not even subjected to any routine testing by the authorities. Under current international safety regulations, a product which contained contaminants as dangerous as those found in the L-tryptophan could still be passed as safe for human consumption.

In 1996, when the UK Advisory Committee on Novel Foods and Processes approved riboflavin (vitamin B2) that had been produced with GE bacteria, they accepted as evidence of safety, data which only identified contaminants present at levels greater than 0.1%.[25] This is clearly inadequate—in the case of L-tryptophan, the level of contamination was far less than 0.1%, and yet proved fatal.[26]

The potential for allergic reactions

In the United States, a quarter of all people report that they have an adverse reaction to one or more foods: most commonly dairy products, eggs, wheat and nuts.[27]

All foods contain proteins, the basic building materials of a cell. For people who are unable to tolerate certain proteins, eating even trace amounts of foods containing them causes allergic reactions, which range from minor discomfort to serious illness and even death. In genetic engineering, genes are transferred from one organism to another. This gene transfer results in the production of new proteins. If a new protein happens to be one that causes an allergic reaction, food that was previously safe for a person could thus become dangerous for him or her to eat.

Pioneer Hi-Bred International, for example, engineered soybeans with a gene from a brazil nut in the hope that it would improve the soybean's protein content. Researchers at the University of Nebraska tested these soybeans on samples of blood serum taken from people who were allergic to brazil nuts. The tests indicated that if these people had eaten the soybeans, they would have suffered an allergic reaction that could have been fatal.[28]

> *"In the special case of transgenic [GE] soybeans, the donor species was known to be allergenic, serum samples from persons allergic to the donor species were available for testing and the product was withdrawn ... The next case could be less ideal, and the public less fortunate."*
> —Marion Nestle, The New England Journal of Medicine[29]

Because most genes being introduced into GE plants come from sources which have never been part of the human diet, there is no way of knowing whether or not the products of these genes will cause allergic reactions.

Some people could develop a sensitivity to a GE food gradually after being exposed to it over time, whereas others might have an acute allergic reaction after eating a minute amount. Unfortunately, the lack of proper labelling or segregation effectively undermines any attempt to monitor GE foods. If allergies do develop, it will be extremely difficult to trace them to their source.

Antibiotic resistance

It is estimated that in the UK, where in some hospitals strains of the common pathogen *Staphylococcus aureus* are resistant to almost all known antibiotics,[30] antibiotic-resistant bacteria kill more people every year than road accidents.[31] The main causes suspected for the build-up of resistant bacteria are the overuse of antibiotics in medicine and animal feed.

A study from East Germany demonstrates the speed at which resistance can spread. In 1982, the antibiotic streptothricin began to be administered to pigs. By 1983, plasmids resistant to streptothricin were found in the pigs' gut bacteria. This resistance had spread to the gut bacteria of farm workers and their family members by 1984, and to the general public and pathogenic strains of bacteria the following year. The antibiotic was withdrawn in 1990, yet the prevalence of the resistant bacteria remained high when monitored in 1993.[32]

The marker genes used in genetic engineering confer resistance to antibiotics commonly used in human and veterinary medicine. Some scientists believe that eating GE food containing these marker genes could encourage gut bacteria to develop antibiotic resistance.

In 1996, for example, the Advisory Committee on Novel Foods and Processes advised the UK government to vote against an authorisation being sought by Novartis for a variety of GE maize containing a marker gene resistant to ampicillin. They felt that the presence of this intact marker gene, together with a bacterial promoter gene that would enable it to operate in bacteria, posed an unacceptable risk.[33]

Experiments are also showing that the potential exists for genes which code for antibiotic resistance (or any other genes) to be transferred to bacteria and other microorganisms from GE crops growing in the field.

In one experiment, genetically engineered rape, black mustard, thorn-apple and sweet peas, all containing antibiotic-resistant genes, were grown together in the laboratory with the fungus *Aspergillus niger*. In some cases their leaves were added to the soil. In each of the experiments, the genes for antibiotic resistance ended up being transferred to the fungus.[34]

Chapter Two

Genetic Engineering and the Environment

Herbicide resistance

Until now, most of the research by the biotech industry has focused on making crops resistant to their own 'broad-spectrum' herbicides. This means that a field can be sprayed with chemicals and nearly all plants will die except the resistant crop. Of the 27.8 million hectares of genetically engineered crops planted worldwide in 1998, 71% were herbicide-resistant.[1] The companies developing these crops are increasing their production capacity for the herbicides, and also requesting permits for higher residues of these chemicals in genetically engineered food.

- Monsanto, for example, has already received permits for a threefold increase in herbicide residues on genetically engineered soybeans in Europe and the United States—up from six parts per million to 20 parts per million.[2]

- Monsanto's Roundup, the world's best-selling weedkiller, accounts for 17% of their total annual sales of $9 billion.[3] Although the US patent on Roundup is due to run out in the year 2000, Monsanto has effectively extended its monopoly: farmers who grow their genetically engineered soybeans sign a contract which opens them to prosecution if they use any herbicide formulations other than the company's own.[4]

- In September 1998, Monsanto announced plans to invest a total of $550 million in a manufacturing and formulation plant for Roundup in Brazil. Previous investments totalling $380 million over the past three years have allowed Monsanto to expand its production of Roundup at facilities around the world.[5]

- In 1997 the British Agrochemical Association predicted that the sales of herbicide in the US would actually benefit from the expanded use of transgenic herbicide-resistant crops.[6]

- AgrEvo, who have increased production facilities for their herbicide glufosinate in the US and Germany, expect sales to increase by $560 million over the next five to seven years. Indeed, "the introduction of glufosinate-resistant crops to increase sales of its herbicide products is considered to be AgrEvo's underlying aim in entering the GE market in the first place."[7] Despite claims that glufosinate is 'environmentally friendly', it is highly toxic to humans and animals, particularly affecting the nervous system. The US Environmental Protection Agency also states that it is toxic at very low concentrations to many aquatic and marine invertebrates. This is particularly worrying because glufosinate is water-soluble and readily leached from soil into groundwater.[8]

- Roundup is also promoted as an environmentally benign herbicide, yet the US Fish and Wildlife Service has identified 74 endangered plant species potentially threatened by excessive use of glyphosate, its principal constituent.[9] Studies have shown that glyphosate can kill fish in concentrations as low as 10 parts per million,[10] that it reduces growth of earthworms and increases their mortality,[11] and that it is toxic to many of the beneficial mycorrhizal fungi which help plants to take up nutrients from soils.[12] It is also the third most commonly reported cause of pesticide-related illness among agricultural workers in California, the only state which produces such statistics.[13] Symptoms include eye and skin irritation, cardiac depression and vomiting.[14]

Margaret Mellon, from the Union of Concerned Scientists, believes that many farmers may be turning towards GE herbicide-resistant crops because they are becoming desperate for new weed control tools. "Farmers in the land of monocultured corn [maize] and soybeans are facing increasingly serious weed problems," she says. "Many weeds have become resistant to chemical herbicides, multiple applications of herbicide have become routine, and new weeds continue to pop up . . . Farmers have also had experience with Roundup around the farm for years and know what a potent plant killer it is. They can easily believe the claims that one application of herbicide will replace several applications of other herbicides."[15] Chuck Benbrook, a former executive of the US National Academy of Sciences Board on Agriculture, agrees: "Weed management is probably the number one management challenge all soybean farmers face . . . Monsanto should not be ashamed to cite these reasons in explaining why the technology is being adopted."[16]

This, together with the fact that Monsanto reduced the price of Roundup by nearly a third in September 1998,[17] certainly seems to make these varieties an economically attractive option to many US farmers. However, as Margaret Mellon points out, these benefits are likely to be short-lived: "Sooner or later weeds will begin to develop resistance to Roundup and more applications of the herbicide will be required. Increasing use of Roundup, of course, will likely increase the rate of herbicide resistance development and pretty soon, farmers will again have lots of weeds and even fewer weed control options."[18]

Repeated applications of a single herbicide encourage plants to develop resistance within a very short period of time: reports from Australia indicate that ryegrass, a commonly occurring weed, may have already developed resistance to Roundup after only 10 sprayings in 15 years. The ryegrass studied survived seven times the herbicide concentration that killed other plants.[19]

Short-term gains, or long-term sustainability?

"Herbicide-tolerant crops perpetuate and extend the chemical pesticide era and its attendant human health and environmental toll. Crops genetically engineered to resist herbicides, insects, and virus diseases, like chemical pesticides, will be sold to farmers as single, simple-to-use products to control pests and sustain continuous monoculture. They are being developed to fit immediately and easily into conventional agriculture's industrialized monoculture. Biotechnology is being developed with the same vision that promoted chemicals to meet the single, short-term goals of enhanced yields and profit margins. This vision embraces a view of the world characterised by beliefs that nature should be dominated, exploited and forced to yield more; by preferences for simple, quick, immediately profitable 'solutions' to complex ecological problems; by 'reductionist' thinking that analyses complex systems like farming in terms of component parts, rather than as an integrated system; and by a conviction that agricultural success means short-term productivity gains, rather than long-term sustainability."—Jane Rissler, Union of Concerned Scientists[20]

As crops and weeds naturally begin to develop resistance, and herbicide-resistant traits are transferred from genetically engineered crops to other plants via cross-pollination (*see pages 36-37*), higher and higher doses of chemicals will be needed to have the same effect, and long-term herbicide use may rise. Alternatively, farmers could turn to crop rotation, innovative cultivation techniques, intercropping, and other methods for weed control. "These methods are harder to adopt than a new variety of soybean, but once adopted, work reliably and safely over the long term," says Mellon.[21]

A number of conservation agencies, such as the Royal Society for the Protection of Birds and English Nature in the UK, have warned about the damage to wildlife that could occur if large areas of land are sprayed with broad-spectrum herbicides such as Roundup. They fear that the increased use of these herbicides will kill the weeds which support the insects and produce the seeds fed on by birds. This could be the final blow for such bird species as the sky-lark, corn bunting and linnet, already in decline due to industrialised farming practices.

> *"The ability to clear fields of all weeds using powerful herbicides which can be sprayed onto GE herbicide-resistant crops will result in farmlands devoid of wildlife and spell disaster for millions of already declining birds and plants."*
> —Graham Wynne, Chief Executive of the Royal Society for the Protection of Birds

Insect-resistant crops

Bacillus thuringiensis (Bt) is a soil bacterium which produces a toxin that is highly valued by organic farmers. These bacteria have been sprayed on crops for more than 50 years as a safe form of biological pest control. Bt targets particular species of insect, such as caterpillars, and the sprays are especially valuable to organic farmers in instances where there is a serious pest infestation.

Crop plants have now been engineered with the gene for the Bt toxin to give them an in-built insecticide; these transgenic 'insect-resistant' crops were grown on 7.7 million hectares worldwide in 1998.[22] In marked contrast to the occasional application of the Bt toxin in organic farming, the transgenic Bt toxin is produced in the plants all the time they are growing. This means that insects are continually exposed to the toxin, and are therefore under constant pressure to develop resistance.

Bt resistance has already been noticed among some insect populations,[23] and the US Environmental Protection Agency has predicted that most target insects could be resistant to Bt within three to five years.[24] In response to the EPA's approval of genetically engineered Bt cotton, maize and potatoes, in February 1999 a lawsuit was filed by the International Federation of Organic Agricultural Movements, the Center for Food Safety and Greenpeace International. The EPA is charged with failing to address concerns that these approvals could lead to the "wanton destruction" of Bt, the "world's most important biological pesticide". "This is just another short term fix that industry is willing to use up," said Jane Rissler, a senior scientist with the Union of Concerned Scientists. The lawsuit demands that EPA cancel registration of all genetically engineered Bt plants, cease any new approvals and immediately perform an environmental impact assessment.[25]

The toxins present in naturally occurring Bt bacteria are only activated by enzymes in the gut of certain insects. The toxin in many of the genetically engineered crops, however, is in a more active form, and may therefore harm a wider range of insects.

- A recent laboratory study in Switzerland found that when lacewings (beneficial insects that prey on crop pests) were fed cornborers raised on Bt maize, the lacewings suffered from disruption to their development and increased mortality. [26]

- Pollinators, such as bees, could be affected in unexpected ways by insect-resistant transgenic crops. One study found that when bees were given sugar solutions containing high levels of the 'protease inhibitors' that gave a transgenic oilseed rape resistance to insects, they subsequently had difficulties learning to distinguish the different smells of flowers. [27]

- In a laboratory experiment at the Scottish Crop Research Institute, it was shown that potatoes that had been engineered to be resistant to insect pests could also harm beneficial insects further up the food chain. Female ladybirds were fed on aphids that had been eating transgenic potatoes, and when compared to ladybirds fed on a normal diet, they laid fewer eggs and lived half as long. [28]

In laboratory experiments at New York University, researchers found that active forms of Bt, like those found in some types of transgenic crops, do not disappear when added to soil, but instead become rapidly bound to soil particles. Unlike the naturally occurring forms of Bt, they are not degraded by microbes, nor do they lose their capacity to kill insects. The accumulation of these toxins, which could be released into the soil as farmers incorporate plant material into the ground after harvest, could represent a serious risk to soil ecosystems. [29]

Virus-resistant crops

Biotech laboratories are now engineering virus genes into plants to make them resistant to viral infections. Transgenic 'virus-resistant' squash and papaya varieties have already been cleared for commercial production in the US, and permits have been granted or are pending for virus-resistant beets, cucumber, lettuce, melon, pepper, potato, sunflower, tomato and watermelon.[30]

There is evidence that existing viruses may pick up viral genes from virus-resistant crops, potentially leading to new strains of plant viruses able to infect a wider range of species.[31] In one laboratory experiment, for example, transgenic virus-resistant tobacco plants were infected with a virus; this virus then recombined with the viral DNA in the genetically engineered plant to form a new viral strain.

In another experiment, plants were infected with a virus that had been disabled so that it lacked the ability to move from cell to cell. It reacquired this ability when similar genes were taken from another virus and inserted into the plants.[32]

It is argued that genetically engineered crops are no more likely to generate new viruses than any plant that has been infected by two or more different viruses at the same time. "You can go into a potato field or a tomato field, or corn [maize] or wheat, where all these viruses are living together," says plant virologist Chuck Niblett. "You don't get new viruses jumping out of these fields and infecting dogs and small children."[33] Transgenic virus-resistant crops, however, will contain viral genes in all their cells all the time they are growing; this, together with the fact that virus-resistant crops are soon to be released over millions of acres, increases the probability that new viruses will be generated.

Pollen from virus-resistant crops may also transfer viral resistance into wild and weedy populations of plant. This could create new weeds from plants which were previously held in check by the viruses to which they have now acquired resistance.

> "It is obviously not going to be possible to cover every eventuality which might result from the release of genetically modified viruses; every virus construct, every host and every ecosystem will probably generate different results."
> —J. S. Cory, Reviews in Medical Virology [34]

Genetic Pollution

> "Imagine the wholesale transfer of genes between totally unrelated species and across all biological boundaries— plant, animal and human—creating thousands of novel life forms in a brief moment of evolutionary time. Then, with clonal propagation, mass-producing countless replicas of these new creations, releasing them into the biosphere to propagate, mutate, proliferate and migrate, colonising the land, water, and air. This is, in fact, the great scientific and commercial experiment underway as we turn the corner into the Biotech Century."—Jeremy Rifkin [35]

Proponents of the technology tend to play down any environmental risks: "There have been 25,000 trials of GM crops in the world now, and not a single incident, or anything dangerous in these releases," says Thomas Joliffe, research and development manager from Adventa Holdings UK. "You would have thought that if it was a dangerous technology there would have been a slip up by now." [36] Yet on scrutiny we find that little, if any, meaningful assessment of ecological impacts has actually been carried out. Margaret Mellon and Jane Rissler from the Union of Concerned Scientists have been monitoring these inspections. "The observations that 'nothing happened' in these . . . tests do not say much," they write. "In many cases, adverse impacts are subtle and would almost never be registered scanning a field . . . the field tests do not provide a track record of safety but a case of 'don't look, don't find'."

As Julie Shepherd from the UK Consumers Association says, "a complacent attitude toward the risks of GM foods and crops arises from an approach which tends to equate 'no evidence' with 'no risk' . . . the complex interactions which take place in the environment and the difficulties of 'managing' or 'controlling' them in the real world are not fully acknowledged by the current risk culture." [37]

The Precautionary Principle

In January 1998, an international group of scientists, government officials, lawyers, labour and environmental activists met in Wisconsin, USA, to define and discuss the precautionary principle. After meeting for two days, the group issued the following consensus statement:

"The release and use of toxic substances, the exploitation of resources, and physical alterations of the environment have had substantial unintended consequences affecting human health and the environment. Some of these concerns are high rates of learning deficiencies, asthma, cancer,

birth defects and species extinctions, along with global climate change, stratospheric ozone depletion and worldwide contamination with toxic substances and nuclear materials.

"We believe existing environmental regulations and other decisions, particularly those based on risk assessment, have failed to protect adequately human health and the environment—the larger system of which humans are but a part.

"We believe there is compelling evidence that damage to humans and the worldwide environment is of such magnitude and seriousness that new principles for conducting human activities are necessary.

"While we realize that human activities may involve hazards, people must proceed more carefully than has been the case in recent history. Corporations, government entities, organizations, communities, scientists and other individuals must adopt a precautionary approach to all human endeavors.

"Therefore, it is necessary to implement the Precautionary Principle: When an activity raises threats of harm to human health or the environment, precautionary measures should be taken even if some cause and effect relationships are not fully established scientifically. In this context the proponent of an activity, rather than the public, should bear the burden of proof.

"The process of applying the Precautionary Principle must be open, informed and democratic and must include potentially affected parties. It must also involve an examination of the full range of alternatives, including no action." [38]

> *"Our current knowledge does not provide us with the means to predict the ecological long-term effects of releasing organisms into the environment. So it is beyond the competence of the scientific system to answer such a question, although precisely this assumed competence is normally the basis for an authoritative appeal."*—Dr. René von Schomberg, Scientific and Technological Options Assessment Report for the European Parliament [39]

Generations that have grown up with DDT, asbestos, PCBs, nuclear energy and BSE are understandably suspicious of official assertions of safety based on a lack of scientific evidence of harm. However, rather than take a precautionary approach, which would demand that the biotech industry prove that genetic engineering is both necessary and safe, most governments seem to view public resistance as something to be conquered. "I think safety is not an issue," said Professor John Beringer, former Chairman of the UK's Advisory Committee on Releases to the Environment, "the issue is communication with the public."

This view is articulated in a proposal written by D. Conning, called 'Biotechnology—Influencing Public Opinion': "It is imperative that accurate and comprehensible information should be made freely available and widely disseminated. The information should state explicitly—there is no risk inherent in the technology itself. Thus, there are foolproof safeguards that modified microorganisms cannot engender diseases in man or the environment; there is no risk that gene transfer would involve the transfer of unidentified DNA that could induce unplanned changes under any circumstances; there is no possibility that modified DNA could gain access to the body of the consumer." [40]

This confidence is not, however, reflected in a willingness

by the biotech industry to accept liability for damage that may be caused by the introduction of genetically engineered organisms. The world's second largest reinsurance company, Swiss Re, has said that the risks of genetic engineering cannot be covered with classical liability insurance models. "For the insurance industry, genetic engineering is potentially one of the most exposed technologies of the future . . . The risk profile of genetic engineering is extremely diversified and very difficult to anticipate. There is no clear conception of the risks accepted, so how can genetic engineering be insured? . . . Today we must accept that the one-sided acceptance of incalculable risks means that any participants in this insurance market run the risk of not only suffering heavy losses, but also of losing control over their exposure." [41]

Once released, the new living organisms made by genetic engineering are able to interact with other forms of life, reproduce, transfer their characteristics and mutate in response to environmental influences. In many cases they can never be recalled or contained. The probability that one or more of these releases could cause serious ecological harm increases all the time as more and more products are approved.

> *"I have the feeling that science has transgressed a barrier that should have remained inviolate . . . you cannot recall a new form of life . . . It will survive you and your children and your children's children. An irreversible attack on the biosphere is something so unheard of, so unthinkable to previous generations, that I could only wish that mine had not been guilty of it."*—Erwin Chargaff, Professor Emeritus of Biochemistry, Columbia University, and discoverer of 'Chargaff's Rules', which laid the scientific foundation for the discovery of the DNA double helix

Using established knowledge of gene movement from conventionally bred crops into wild plant populations,[42] three scientists from Michigan State University writing in the journal *Hortscience* concluded that gene transfer can be expected to occur "regularly . . . from most if not all transgenic crops". They went on to state that it is not so much a case of whether gene transfer will occur, but a question of what the effects will be when it happens.[43]

In a number of different field trials, genes from oilseed rape that had been genetically engineered to be resistant to glufosinate, a broad-spectrum herbicide, crossed to weedy species, producing fertile, weed-like plants after just two generations of hybridisation and backcrossing.[44] Research in Germany has shown that that these genes can be transferred to crops in fields 200 metres away.[45] Evidence also shows that resulting hybrids will not necessarily die out quickly, as is often suggested. Allison Snow, an ecologist from Ohio State University, found that wild plants containing a gene for herbicide resistance produced just as many seeds and were able hold their own in competition with their unmodified counterparts.[46]

Scientists at the University of Chicago demonstrated that GE plants examined in field tests had a dramatically increased ability to outcross and transfer genes to non-GE plants. The scientists compared rates of gene flow from two different kinds of herbicide-resistant mustard plant, one of which was genetically engineered and the other produced by traditional breeding. Even though both varieties of mustard plant contained the same gene for herbicide resistance, the scientists found that GE plants were 20 times more likely to interbreed with related species when compared to traditionally bred plants. They have not been able to discover why the engineering process appears to alter the rates of outcrossing.[47]

These studies suggest that where there are weedy (and non-weedy) species of plant related to transgenic crops, there could be a rapid transfer of modified genes between the two. This is likely to occur even more frequently in tropical countries: because they are the centres of origin for many of the plants we now consume as food, most of the food crops grown have numerous related species in the wild. In Guatemala, for example, where genetically engineered 'FlavrSavr' tomatoes were recently grown without the consent or knowledge of the authorities, there are hundreds if not thousands of indigenous varieties of tomato.[48]

Crop plants are now being genetically engineered to produce pharmaceuticals and industrial chemicals. These plants could cross-pollinate with related species and contaminate the food supply, and could expose foraging animals, insects and seed-eating birds to a wide range of drugs, vaccines and chemicals.[49] Unfortunately, in the 1995 joint consultation between the World Health Organisation and Food & Agriculture Organisation on the food safety issues raised by genetic engineering, the potential dangers posed by these crops were judged to be "unrelated to food safety", and therefore outside the remit of the consultation.[50]

> "An ecosystem—you can always intervene and change something in it, but there's no way of knowing what all the downstream effects will be or how it might affect the environment. We have such a miserably poor understanding of how the organism develops from its DNA that I would be surprised if we don't get one rude shock after another."—Professor Richard Lewontin, Professor of Genetics, Harvard University [51]

Of the funds it allocates to biotechnology research, by 1997 the US Department of Agriculture was only devoting 1% to risk assessment: a total of only $1-$2 million a year to study the entire range of environmental issues involved in the release of transgenic organisms.[52] The environmental consequences of increasingly frequent releases of genetically engineered bacteria, animals, insects and microorganisms have barely begun to be studied.[53]

- GE mites, mosquitoes and nematodes have been created in laboratories for a variety of purposes.[54] These creatures reproduce quickly and, if released into the environment, would be hard to contain in any specified area.

- Microorganisms have rapid rates of reproduction, readily exchange genetic information, and are very difficult to detect in the environment. Once released into the environment, transgenic microorganisms could create millions of offspring within days or even hours.[55] In 1989, a company called Biotechnica International conducted field trials with soybeans which had been coated with a genetically engineered microorganism in the hope that this would increase nitrogen fixation. At the end of the season the plants and seeds were incinerated, the fields were ploughed under and a new crop was planted. Subsequent monitoring showed that the genetically engineered microorganisms were out-competing microorganisms that normally lived in the soil.[56]

- In 1998, Frank Gebhard and Kornelia Smalla published laboratory studies which demonstrated that gene transfer could occur from GE sugar beet to commonly occurring soil bacteria called *Acenitobacter*. It is possible that insects, birds or other animals could pick up bacteria from the soil and transfer them wherever they go, potentially opening up another route for gene transfer from genetically engineered crops.[57] In another experiment, Gebhard and Smalla found that genes which had been engineered into sugar beet were still present in the soil two years after the beet had been shredded and ploughed back into the earth.[58]

In 1994, scientists based at Oregon State University devised experiments to assess the potential ecological consequences of the release of a genetically engineered microorganism called *Klebsiella planticola*. The bacterium, which was also being reviewed by the US Environmental Protection Agency and other universities in the US, had been engineered to produce increased ethanol concentrations in fermentors that are designed to convert agricultural wastes into an alternative fuel source. In this system, the residues from the fermentation, including the engineered bacteria, would be used as a soil amendment.

In the laboratory, the researchers found that when the genetically engineered *Klebsiella* was added to a small microcosm consisting of wheat plants and sandy soils, it killed the plants, whereas the addition of a *Klebsiella* that had not been engineered did not. The experiments also showed that the bacteria were able to persist in the soil. These results raised the possibility that soil amendments containing the genetically engineered *Klebsiella* could kill or impair crops and beneficial soil life in fields with similar soils and, furthermore, that once released the *Klebsiella* could become established and be very difficult to eliminate.

"When the data first started coming in", said Professor Elaine Ingham, "the EPA charged that we couldn't have performed the research correctly. They went through everything with a fine toothcomb and they couldn't find anything wrong with the experimental design—but they tried as hard as they could. At that time, some EPA researchers did not understand 'ecologically-based' testing systems designed to look at the microbial interactions and nutrient cycling processes that would occur if *Klebsiella planticola* were released into the environment. The fact is that the regulatory system as it stands is totally inadequate to catch these kinds of unexpected effects. If we hadn't done this research, the *Klebsiella* would have passed the approval process for commercial release.

"Before these experiments I used to receive a significant amount of funding from the EPA but now I have given up handing in proposals because they just aren't interested anymore. I don't know how I can prove anything but it does seem like a big coincidence. Life in my department was certainly made difficult for me, and in the end I had to transfer. Independent research can be quite problematic nowadays because most departments receive well over 50% of their funding from private corporations and there can be fallout if you get results that they don't like." [59]

Transgenic fish

About 50 laboratories around the world have been conducting research into transgenic fish.[60] Genes from chicken, humans, cattle and rats have been engineered into commercially important varieties such as carp, catfish, trout and salmon, to increase rates of growth and reproduction.[61]

- In one trial, for example, Atlantic salmon were engineered with growth hormones to enable them to reach adult size more quickly. After one year, most had a two- to six-fold increase in growth, while the largest was 13 times normal size.[62]

- The arctic flounder is a fish that can survive in freezing temperatures. Genes that code for antifreeze chemicals have been taken from the flounder and engineered into salmon in an attempt to enable them to tolerate colder water. If this were successful it would allow fish farming to expand into more northerly regions.

- Pacific salmon are being genetically engineered so that they are able to live and breed in the ocean, rather than follow their annual migrations from salt water to fresh water. If this trait were to pass into wild salmon populations, they would no longer need to return to their native streams to spawn. This could cause severe ecological disruption to river life, and to other species such as bears, which for thousands of years have depended on the annual salmon migrations.[63]

Escapes from fish farms are already a serious problem; in some parts of Norway, fish that have escaped from farms have bred and now outnumber wild ones by five to one.[64] Traits such as increased cold tolerance or faster breeding cycles could give transgenic fish a definite competitive advantage, and if they escaped they could end up displacing local fish populations.

Trees

Many species of tree are also being genetically engineered. Some are designed to be able to resist herbicides, viruses, insects, frost, and drought; others to grow faster and to produce wood that is easier to pulp or to process into lumber.[65] Scientists are also considering the production of a genetically engineered enzyme that could clear up the effluent from paper mills by destroying lignin, the organic substance that makes wood rigid.[66]

As with agricultural products, research on the possible environmental consequences lags far behind the development of the engineered varieties.[67] As we know from experience with some exotic species of tree which were introduced into ornamental gardens before the beginning of this century, but have only recently spread into the wild, it can take well over 100 years for the ecological effects of introducing new types of tree to become apparent.[68]

According to The World Wide Fund for Nature, which in November 1999 called on governments worldwide to declare a global moratorium on GE trees, their commercial production could begin by 2002, probably in Chile, China and Indonesia.[69]

Estimated area of land sown commercially with transgenic crops (millions of hectares) 1997-8*		
Country	1997	1998
USA	8.1	20.5
Argentina	1.4	4.3
Canada	1.3	2.8
Australia	0.1	0.1
Mexico	< 0.1	0.1
Spain	0	< 0.015
France	0	<0.001
South Africa	0	<0.1
Total	11.0	27.8

*Excluding China, for which accurate figures are unavailable.[70]

Chapter Three

Genetic Engineering and Farming

"*The worrying thing is that most people seem to think farmers are pushing forward the genetic agenda rather than the big agrochemical corporations.*"—Cornish Farmer Michael Hart, during his tour of south-west England with the 'Keep Britain Farming' roadshow.[1]

Much of the promotion of genetic engineering has centred on future benefits, but the dreams of high yields or increased nitrogen fixation may in fact be unrealistic, because they involve complex multigene traits. Nitrogen fixing, for example, depends on at least 17 genes in the bacterium and 50 genes in the plant.[2] There are hazards associated with the transfer of a single gene, let alone 50. Even if all the genes required for these traits could actually be identified and transferred, problems of genetic instability could increase as a result.[3] Instability in GE crop lines has already led to crop failures, which have not been well reported.

- In 1997, crop failure affected 30,000 acres of GE herbicide-resistant cotton in Mississippi. Some growers faced losses of between $500,000 and $1 million.[4] Monsanto, producer of the Roundup Ready cotton, reportedly paid out millions of dollars in out-of-court settlements. In 1998, the Mississippi Seed Arbitration Council ruled that

Monsanto's cotton failed to perform as advertised, and recommended payments of nearly $2 million to three cotton farmers who suffered severe losses.[5]

State and Federal cotton experts say that the company hurried the genetically engineered cotton onto the market without letting them test it. Bill Meredith, a geneticist and research manager for the US Department of Agriculture in Mississippi, asked Monsanto for a single pound of cotton seed, enough to test on just a tenth of an acre. He was told that Monsanto couldn't spare that much, even though farmers were planting the cotton on thousands of acres. "We weren't able to find out what was going on," he said. "These new varieties and new technologies are going out with less evaluation than they had in the past with traditional varieties."[6]

- In 1996 Monsanto's 'New Leaf' potatoes, genetically engineered with the Bt toxin, were planted in three regions of Georgia in the former Soviet Union. Farmers suffered yield losses of up to two-thirds of their entire crops. It is thought that the potatoes may have been affected by a disease called phytophtora because they were not suited to the local conditions. Many farmers were pushed into debt as a result.[7]

- In 1994, Calgene (now a subsidiary of Monsanto) introduced the 'FlavrSavr' tomato, the first genetically engineered whole food approved for commercial sale. It was engineered to ripen longer on the vine and still be hard enough to withstand the processes of picking, packing, and transport. By 1997 it had been withdrawn from the market. Contrary to Calgene's expectations, the tomatoes were often so soft and bruised that they could not be sold as fresh produce. Also it was found that most of the FlavrSavr varieties did not have acceptable yields or disease resistance.[8]

- Monsanto's Bt cotton planted in 1996 was supposed to be resistant to the bollworm. Instead, nearly half of about two million acres of Bt cotton in the southern United States suffered a heavy infestation, and growers were advised to salvage the crop with emergency spraying. In spite of claims that the Bt cotton would be 90-95% effective, some cotton consultants reported that the product was only 60% effective. A legal firm in Texas acting for 17 of the growers claims that Monsanto misrepresented the product.[9]

According to Monsanto, in 1998 their Roundup resistant soybeans averaged 43.1 bushels per acre, beating conventional growers by 4.5 bushels. A number of other studies, however, do not support these conclusions. Ed Oplinger, for example, Professor of Agronomy at the University of Wisconsin, has been conducting performance trials for soybean varieties for the past 25 years. His comparison of yields in the 12 states that grow 80% of the soybeans in the United States show that, on average, the yields of genetically engineered soybeans were 4% lower than conventional varieties.[10]

In another study, researchers Marc Lappé and Britt Bailey compared the performance of Monsanto's Roundup-resistant soybeans with those of conventional varieties grown under the same conditions. Taking their data from the 1996 figures assembled by the Arkansas Cooperative Extension Service, they matched the transgenic soybeans to their nearest conventional type and compared the number of bushels per acre harvested at the end of the 1995 and

1996 growing seasons. In 30 out of 38 varieties, the conventional soybeans outperformed the transgenic ones, with an overall drop in yield among the transgenic soybeans of an average 4.34 bushels per acre. This amounts to a loss of nearly 10% when compared to conventional varieties.[11]

Marc Lappé and Britt Bailey also visited a number of farmers and agricultural scientists in the south-east United States in 1997, to ask them about their experiences with the new herbicide-resistant crops. One of them, Dr. Ford Baldwin, is an extension agent and weed scientist from the University of Kansas, who works with approximately 250 farmers throughout the state. Dr. Baldwin told them that he had received more complaints about pesticide drift that year than any of the 20 years that he had been in the field. He believes this is largely due to the mass use of Roundup-resistant crops and concurrent increase in the acreage that is being sprayed. "Pesticide-drift caused crop destruction increases the pressure for non-users . . . to get on the Roundup Ready bandwagon," he remarked. "A neighbouring farmer not using the technology stands to have his fields destroyed if the Roundup herbicide drifts onto his fields."[12]

Bill Christison, President of the US National Family Farm Coalition, who operates a 2,000-acre farm producing soybeans, maize, wheat, hay and cattle, is another farmer who is unhappy with the genetically engineered soybeans. He quotes statistics from yield books in Missouri showing that GE soya had a five bushel per acre decrease in yield compared to conventional soya.

> "The promise was that you could use less chemicals and produce a greater yield. But let me tell you none of this is true . . . they wanted us to sign a production contract which

limited what we could do with our production. It is our practice to produce our own seed for the following year's planting. Because the contract forbids this, it would have cost us three times as much for seed. And then there's a problem of paying for a patenting fee of several dollars per bag . . . Then, we found GMO seed actually produces a lower yield because of the varieties that had been altered. The acceptance of GMOs by the US farmer is predicated by the fact that farmers are hard pressed to survive financially, and have become acclimated to the idea that new technology is good technology." [13]

Pinkerton private detective agency, which used to supply employers with auxiliaries to break trade unions, has been hired by Monsanto to check that farmers are not saving seeds.[14] This age-old practice is forbidden under the 'Technology Use Agreement', which also allows the company to access farmers' land to take plant samples during a three-year period after the genetically engineered seed has been purchased.[15] A freephone hotline has also been set up to encourage farmers to tell on their neighbours for seed-saving. By late 1998, more than 475 farmers in the United States and Canada had already been sued by Monsanto or are awaiting lawsuits for breaking their contracts.[16] It has been company policy to broadcast radio advertisements in which they name farmers who have been caught saving seed.[17]

One of the US farmers who was 'named and shamed' in such an advertisement reportedly had to pay the company US $35,000 damages and sign an agreement to abstain from criticising Monsanto.[18] Another farmer from Saskatchewan in Canada is being sued for growing seed without a licence, after samples were taken from around his fields. Percy Schmeiser says that he has been growing oilseed rape for years and freely admits to saving his seed, but denies that it belonged to Monsanto. The problem, according to Mr. Schmeiser, is that there are a lot of genetically engineered crops being grown in the neighbouring area and pollen from them is blowing everywhere. "It's in the ditches and the roadsides; it's in the shelterbelts; it's in the gardens; it's all over . . . We're just touching the tip of the iceberg in contamination of fields by this Roundup genetic canola [oilseed rape] . . . It just opens up a vast area of uncertainty." [19]

Cross-pollination

"The Government is absolutely committed to making sure that those who do not want to eat crops that have been cross-contaminated, or to have their crops cross-contaminated, have their rights in this protected as well."—Nick Brown, UK Agriculture Minister [20]

In spite of such assertions, governments have failed to protect the interests of farmers concerned that genetically engineered crops planted nearby could cross-pollinate their own. In July 1998 an organic farmer, Guy Watson, appealed to the UK government not to go ahead with a trial of GE maize that was due to be planted 275 metres from the farm where he grows his organic sweetcorn. The Soil Association, the largest organic certification body in the UK, told Guy Watson that they would be forced to revoke the organic status of his sweetcorn if there was found to be any evidence of cross-pollination from the GE maize. The government refused to call a halt to the trial after the government's Advisory Committee on Releases to the Environment (ACRE) decided that the degree of cross-pollination of the organic sweetcorn would be no more than 1 kernel in 40,000 at 200 metres. Minutes taken during ACRE's meeting noted that "some members thought this proportion was too high." [21]

In January 1999 the Soil Association commissioned an independent report by the National Pollen Research Unit (NPRU) at the University of Worcester.[22] Dr. Jean Emberlin, having looked at all the research available, estimated that "in conditions of moderate wind speeds the rates of cross-pollination at 200 metres would be in the order of one kernel in 93"—an average of about five kernels in each cob of sweetcorn. The report also included references showing that bees pick up pollen from maize plants and can carry it for several miles—something that was not even considered by ACRE, despite the fact that there were 20 bee hives on the border of the trial site. The report concluded that "overall it is clear that the maize pollen spreads far beyond the 200 metres cited in several reports as being an acceptable separation distance to prevent cross-pollination."

Another report published in April 1999 supports this conclusion. Scientists in the UK planted male-sterile oilseed rape plants at various distances up to 4,000m from a field in which transgenic oilseed rape was being grown. The researchers used male-sterile plants which are not able to self-pollinate so that they would know for sure that any seeds produced must come as a result of cross-pollination from the field. The scientists found that even at 4,000 metres, 5% of flower buds on the test plants were pollinated.[23]

- 87,000 packs of organic tortilla chips worth over £100,000 were recalled and destroyed in the UK after a routine analysis revealed that transgenic DNA was present in the product. The maize used had been grown on a 7,000-acre organic farm in Texas, a region in which many farmers grow GE maize, and after investigation the suppliers concluded that the most likely source of the contamination was cross-pollination.[24] It has not yet been established whether farmers growing transgenic crops will be held liable in cases such as this.

Premiums for GE-free crops

High premiums are now being paid for produce that can be guaranteed GE-free. In such a market, farmers who refuse to grow GE crops have a unique opportunity to compete with cheaper imports.

In January 1998, Australian trade authorities announced the largest shipment of canola (oilseed rape) ever exported from Australia, bound for processing plants in Europe. According to Graham Lawrence, Managing Director of the New South Wales Grains Board, "Europe has moved to become a major buyer this year because Australia is the only country to guarantee non-genetic modified canola." Canada, on the other hand, lost $300-400 million in oilseed rape sales to Europe in 1998, because 50% of the crop had been genetically engineered. Government authorities failed to ensure segregation of GE and non-GE grains, and the Canadian exporters were unable to guarantee GE-free status to meet the demand.[25]

In a recent report to the British government, the Royal Institute of Chartered Surveyors (RICS), who have over 74,000 members, said that the growing of genetically engineered crops could reduce the value of agricultural land and potentially leave farmers open to legal action. The report suggested that a land register should be set up, which potential buyers and banks could consult to find out if and when GE crops had been grown on a particular holding, noting that "the BSE crisis has shown that the lack of adequate records hampered the industry's attempt to win back consumer confidence." The RICS also said that the presence of transgenic crops could become as relevant to purchasing a piece of land as any past contamination, location close to slag heaps or a history of crop disease. In response to the report, the National Farmers Union of Scotland said that "anybody who was asked to grow GM crops today would say no because it would be commercial suicide." [26]

Genetic engineering and biodiversity

Some releases of genetically engineered organisms pose the same risks to biodiversity as the introduction of non-native species into new habitats. There are, for example, plans to genetically engineer major crops such as rice and wheat with new traits, such as increased salt tolerance. This may enable them to be grown on land which would previously have been regarded as unsuitable—either left wild or used for growing indigenous crop plants adapted to local conditions. These traits could give transgenic crops a competitive advantage over native plants, potentially causing serious ecological disruption.

It is even more likely that disruption could occur if cross-pollination were to transfer advantageous traits to wild plants, which could then become more vigorous. "Even rare genetic transfers to wild plants could have devastating effects," says ecological geneticist Norm Ellstrand of the University of California. He is concerned that, even if such an event happened far less than 1% of the time, "within 10 years we will have a moderate to large scale ecological or economic catastrophe, because there are so many products being released." [27] In the United States, 42% of the species on the threatened or endangered species list are at risk primarily because of non-indigenous species. [28] According to the US Department of the Interior and researchers at Cornell University, invasive non-indigenous species (including diseases) cost the US economy an estimated $123 billion a year. [29]

- The damage that can be caused by the introduction of even a single new species is illustrated by the example of 'European purple loosestrife', an ornamental plant that was introduced to the US in the early 19th century. [30] This plant has been spreading at a rate of 115,000 hectares per year and is changing the basic structure of most of the invaded wetlands in the United States. [31] It now occurs in 48 states, and costs $45 million per year in control measures and forage losses. [32]

"If we can take animals whose characteristics are well-known, well-understood, and reasonably predictable and put them into environments that are familiar, and we still occasion disaster—sometimes disaster that we can't reverse—how much more likely are we to do so with new organisms, whose traits we do not yet understand?"—Bernard Rollin [33]

With widespread deforestation, pollution, and habitat destruction, it is estimated that at least 30,000 species worldwide are becoming extinct every year. [34] Furthermore, according to the UN Food and Agricultural Organisation, we have lost 75% of the genetic diversity present in agriculture at the beginning of this century, principally as a result of industrialised farming practices. [35] This trend is likely to continue with GE crops, because they are designed to fit into the very systems of monoculture which have proved to be so destructive.

> *"Although biotechnology has the capacity to create a greater variety of commercial plants, the trend set forth by Transnational Corporations is to create broad international markets for a single product, thus creating the conditions for genetic uniformity in rural landscapes."*—Miguel Altieri[36]

Genetic uniformity leads to vulnerability.[37] In the case of the Irish potato famine in the 19th century, for example, genetic uniformity in the potato crop meant that all the potatoes were susceptible to a single disease.[38] The same potato blight also struck the Andes, but there farmers plant as many as 46 varieties of potatoes, and this genetic diversity gave them protection—the disease only affected a few varieties. Potatoes from the Andes were subsequently used to restock the European farms.[39]

Biodiversity is traditionally understood to be the very basis of food security. The more genetic diversity there is within an agricultural system, the more that system is able to accommodate challenges from pests, disease or climatic conditions which usually only affect certain varieties. For millennia, farmers around the world have been using organic farming techniques to protect crops from pests, fungal and viral infections. These include highly sophisticated systems of multiple-cropping.

- Mexico's Huastec indian communities have highly developed forms of forest management in which they cultivate over 300 different plants in a mixture of small gardens, agricultural fields and forest plots.[40]

- One village in north-east India grows up to 70 different varieties of rice.[41] In West Bengal, 124 'weed' species collected from rice fields have economic importance for farmers.[42]

- In the Expana region of Mexico, farmers make use of 435 wild plant and animal species, of which 229 are eaten.[43]

The Green Revolution

Parallels can be drawn between the 'gene revolution' and the 'Green Revolution'. The 'Green Revolution' was a massive government and corporate campaign that persuaded farmers in the Third World to replace a multitude of indigenous crops with a few high-yielding varieties, dependent on expensive inputs of chemicals and fertilisers. This led to huge losses in genetic diversity. In 1996 an FAO report listed the main causes of plant genetic erosion in 154 countries: in over 80 of them, replacement of local varieties came top.[44] Many of the indigenous varieties that farmers used to grow have now been lost for ever.

> "A few decades ago, Indian farmers were growing some 50,000 different rices; just over ten years ago, this number had dropped to 17,000; and today, the majority grow just a few dozen. In Indonesia, 1,500 local varieties have become extinct in the last 15 years. If different varieties, each of which have different traits, are not grown out constantly, they are very quickly lost." —Corner House Briefing [45]

The insecticides and herbicides that went hand in hand with the use of Green Revolution crops caused the loss of complementary harvests which had previously been provided by the paddy fields, such as fish, shrimp, crabs, edible herbs, frogs and wild plants. The loss of these harvests is seldom taken into account when yields of Green Revolution—

or genetically-engineered—crops are calculated.[46]

- In the late 1970s, vast areas of land were planted in Indonesia with a single variety of rice. These crops, which had been sprayed with pesticides, were devastated by a pest called the brown plant hopper,[47] leading to severe food shortages. Subsequent monitoring has shown that in fields just a few metres away, where pesticides have not been used, predators of the plant hopper flourish and healthy rice plants continue to grow.[48]

- Since pesticides were introduced in the US in the 1940s, the proportion of crops lost to insects has grown by 13%.[49]

Green Revolution crops have also led to extremely high levels of pesticide-related illness. In 1990 the World Health Organisation estimated that occupational pesticide poisonings may affect 25 million people worldwide each year, including 3 million severe poisonings and 220,000 fatalities.[50] It is estimated that 99% of these fatalities occur in Third World countries,[51] where multinational corporations frequently market pesticides that are banned in industrialised countries.[52]

Feeding the world

Genetic engineering, we are told, is now essential to meet the needs of a world population that is increasing by 88 million people every year.[53] According to this view, it is selfish and shortsighted to suggest otherwise. As Monsanto said in their 1998 advertising campaign, used to promote genetic engineering in Europe: "Slowing its acceptance is a luxury our hungry world cannot afford."

In June of that year, as these advertisements were appearing in British newspapers, there was a meeting at the UN Food and Agricultural Organisation on the issue of plant genetic resources. Twenty-four delegates from 18 African countries, who were representing their governments at this meeting, decided to respond to the adverts with a statement to the press: "We ... strongly object that the image of the poor and hungry from our countries is being used by giant multinational corporations to push a technology that is neither safe, environmentally friendly, nor economically beneficial to us. We do not believe that such companies or gene technologies will help our farmers to produce the food that is needed in the 21st century. On the contrary, we think it will destroy the diversity, the local knowledge and the sustainable agricultural systems that our farmers have developed for millennia and that it will thus undermine our capacity to feed ourselves." [54]

Even if genetic engineering were able to deliver its promises of high-yielding, disease-resistant crops for the Third World, it seems unlikely that this would be of benefit to starving populations because it fails to address the root causes of hunger. Indeed, the suggestion that this complex problem can be solved with a biotechnological panacea allows both governments and industry to distance themselves from their complicity in the political structures and social inequalities that lead to starvation. According to the United Nations' World Food Programme, we are already producing one and a half times the amount of food needed to provide everyone in the world with an adequate and nutritious diet; yet one in seven people is suffering from hunger.

Responding to a British scientist's claim that those who want to ban genetically engineered crops are undermining the position of starving people in Ethiopia, Tewolde Berhan

Gebre Egziabher, Spokesperson for the African Group at the FAO, and General Manager of the Environmental Protection Authority in Ethiopia, remarked: "There are still hungry people in Ethiopia, but they are hungry because they have no money, no longer because there is no food to buy . . . We strongly resent the abuse of our poverty to sway the interests of the European public." [55]

- In 1997, the UN Development Report stated: "In Africa alone, the money spent on annual debt repayments could be used to save the lives of about 21 million children by the year 2000." World Bank and OECD statistics from 1997 show that for every $1 that the West gave in aid to the poorest countries in the world, these same countries paid back $6.32 to the West in debt, a staggering $836.2 million every day. Of the total figure of debt repayments by Third World countries that year—$305.2 billion—interest charges ($109.1 bn) made up more than a third. [56]

- At the height of the 1984 famine in Ethiopia, oilseed rape, linseed, and cottonseed was being grown on prime agricultural land to be exported as feed for livestock to the UK and other European countries. [57] Other products exported to Europe from Ethiopia during the famine included coffee, meat, fruit and vegetables. [58]

- In South America, per capita food supplies rose by 8% between 1970 and 1990, but the number of hungry people went up by 19%. [59]

The problem is not, however, just one of distribution. With the massive increases in population that we are witnessing, there is a corresponding decrease in the amount of agricultural land available. Hundreds of thousands of hectares are being paved over every year by urban sprawl and industrial growth—in the United States alone for example, some 168,000 hectares (an area equivalent to twice that of New York City) was paved over every year between 1982-92.[60] Another reason for loss of land is degradation of the soil due to erosion, contamination or compaction. In certain cases this is so severe that land is taken out of production—estimates vary between 50,000 and 100,000 square km every year.[61]

Given the problems outlined above, we urgently need to explore ways of producing food which meet a range of social, political and agricultural challenges. Ecologically based agricultural systems, as practised in many parts of the world, aim to address these critical issues by emphasising:

- Local production of food adapted to and integrated with its ecological and socioeconomic setting.

- Diversity—through intercropping (planting different crops together); crop rotations; and the cultivation and preservation of biodiversity.

- Reduction in nutrient losses—effectively containing leaching, runoff and soil erosion, and minimising soil degradation—by maintaining vegetative cover, disturbing the soil as little as possible, and recycling resources.

- Improved nutrient recycling and preservation of natural resources—through the use of cover crops, organic matter such as manures and composts, and the promotion of a healthy soil.[62]

These principles have been successfully applied in Cuba. Until 1989 the country imported nearly half its food, and its agricultural landscape was dominated by large scale industrialised farms, dependent upon imported machinery and agrochemicals. With the end of the Soviet Union, however, and the subsequent collapse of trade, together with a continued US trade embargo which made it very difficult for Cuba to import food and agrochemicals, the country was plunged into a food crisis. Forced to become self-reliant, the government turned towards alternative agricultural systems based on small-scale, mainly organic agriculture, local food production, and land reform, which included innovative urban farming initiatives. Cuba's farms are now hailed as a great success, and proof that it is possible to develop new models of agriculture without the use of chemicals, and still feed a growing population.[63]

In the US, long-term studies of organic farming methods have also produced encouraging results:

"Our results from the first 14 years show that comparable yields can be obtained without the use of chemical pesticides or fertilizers. Corn [maize], which is the most demanding crop in terms of nutrients, suffered a 25% yield reduction during the first four years but had yields comparable to the conventional system in subsequent years. Yields are highly variable from year-to-year because of climatic variability. During drought years, all corn yields are reduced because of insufficient water. However, we have noticed that yields in the organic corn are not reduced as much as the conventional corn, suggesting that the organic systems may be more resistant to drought-stress."—Rodale Institute Farming Systems Trial 1981-95 [64]

Other examples have been documented by Jules Pretty, Director of the Centre for Environment and Society at the University of Essex in the UK, who has done research into the application of sustainable farming practices in the Third World. He found that:

- 223,000 farmers in southern Brazil have used green manures and cover crops of legumes together with livestock integration. They have doubled yields of maize and wheat to four to five tonnes per hectare.

- 45,000 farmers from Guatemala and Honduras have used regenerative technologies to triple maize yields to 2-2.5 tonnes per hectare and diversify their upland farms. This has led to local economic growth that has in turn encouraged re-migration from the cities.

- More than 300,000 farmers in southern and western India, farming in dry-land conditions, are now using a range of water and soil management technologies. They have tripled sorghum and millet yields to 2-2.5 tonnes per hectare.

- Approximately one million small coffee farmers in Mexico have adopted fully organic production methods, increasing yields by half.

- 200,000 farmers across Kenya have taken part in various government and non-government soil and water conservation and sustainable agriculture programmes, which have been doubling their maize yields to about 2.5-3.3 tonnes per hectare and substantially improved their vegetable production during the dry seasons.

- Yields have been falling in high-input rice varieties introduced during the Green Revolution. However, over a million wetland rice farmers in Bangladesh, China, India, Indonesia, Malaysia, Sri Lanka, Thailand, Vietnam and the Philippines are now increasing their yields without the use of chemicals by adopting sustainable agricultural practices.[65]

These sustainable agricultural systems are able to provide substantial increases in yields whilst encouraging the use of local resources and helping communities to become more self-reliant. In contrast, multinational corporations, who are in the business of selling seeds, fertilisers and chemicals, aim to tie farmers to external inputs—which come only from them, at their price. Such corporations are naturally reluctant to acknowledge the potential of agricultural systems that are outside their control.

The assumption that we need to create new crop varieties through the use of genetic engineering technologies overlooks the fact that there is untapped potential within the wealth of existing varieties. A report for the US National Research Council, for example, drew attention to traditional African cereals—in particular to their potential for expanding and diversifying African and world food supplies. "What has been almost entirely overlooked," says the report, "is that throughout that vast continent can be found more than 2,000 native grains, roots, fruits and other food plants. These have been feeding people for thousands of years but most are being given no attention whatever today."[66]

Productivity of small farms

The 1980 World Census on Agriculture carried out by the UN Food and Agricultural Organisation, showed that when compared acre to acre with large farms, smaller farms, which tend to have greater diversity and are usually more carefully cultivated, were consistently found to be more productive.

- Farms of between one and two and a half acres in Syria were, on average, over three times more productive than farms of 37 acres.

- Farms of between one and two and a half acres in Nigeria were, on average, over four times more productive than farms of 37 acres.

- 7-10 acre farms in Mexico were, on average, over 12 times more productive than farms of 37 acres.

- Farms of between 12 and 17 acres in Peru were, on average, over four times more productive than farms of 37 acres.[67]

Land reform

"Landlessness and hunger go hand-in-hand. Eight out of ten farmers in Central America do not have enough land to sustain their families, forcing them to look for seasonal jobs. In Guatemala, huge swathes of land owned by the biggest landlords—an estimated 1.2 million hectares—lie idle, either because the price of export crops is too low to justify planting or because the land is being held for speculation. Some 310,000 people over 20 years of age are landless and without permanent employment." —Corner House Briefing , 'Genetic Engineering and World Hunger: Greed or Need?'[68]

Comprehensive land reforms in Japan, Zimbabwe, and Taiwan have markedly increased food production. In a World Bank study carried out in north-east Brazil, it was estimated that redistributing farmland into smaller units would increase the output by 80%.[69]

Land usage could also be reformed by giving higher priority to essential food crops. The five million hectares currently being used to grow tobacco could alternatively produce 15 million tons of grain.[70] Another example is the use of soybeans; if we ate the beans ourselves, rather than feeding them to livestock (which is how 90-95% are currently used), we would get far more protein per acre. One steer eats over 790 kg of plant protein, but produces less than 50 kg of meat protein, most of the energy it receives from the plant being used to maintain normal bodily functions.[71]

Debt

When industrialised farming systems, which were originally designed to maximise the productivity of labour, were exported to countries where labour was abundant but capital scarce, they caused enormous social upheaval and rural-urban migration. In South Korea, for example, following the implementation of the Green Revolution, the number of rural households in debt rose from 76% in 1971 to 98% in 1985. In the Punjab, the high costs of chemicals and fertilizers for Green Revolution crops led to a decline in the number of small farms by nearly a quarter between 1970 and 1980.[72]

'Microcredit' schemes have been developed successfully in a number of countries to help people caught in the poverty trap by providing them with small loans at very low interest. Multinationals have noted their success, and are beginning to introduce their own microcredit systems, but often with higher rates of interest and very different goals. As Monsanto's Chief Executive, Bob Shapiro, recently told readers of the house magazine of the International Finance Corporation (part of the World Bank which focuses on private investment in developing countries): "It is truly easy to make a great deal of money dealing with very primary needs: food, shelter, clothing."[73]

In 1998 Monsanto tried to set up a joint venture with the Grameen Bank, a highly regarded microcredit scheme in

Bangladesh which provides credit to the poor. The corporation intended to call it 'The Grameen-Monsanto Centre for Environmentally Friendly Technologies'. Suspicious that the real motivation behind the deal was to hook the country's poorest farmers onto expensive Monsanto products, Grameen pulled out. Undeterred, Monsanto reported that it was "striving to have at least one microcredit project operating in each of its world areas by the end of 1998", "working with third-party organisations in Indonesia, India and Mexico", and "also in the early stages of investigating partnerships in such areas as Eastern Europe, China, South Africa, sub-Saharan Africa and parts of Latin America." [74]

> *"Rather than reducing world hunger, genetic engineering is likely to exacerbate it. Farmers will be caught in a vicious circle, increasingly dependent on a small number of giant multinationals, such as Monsanto, for their survival . . . The truth is that genetically engineered crops will provide a 'better way forward' for Monsanto's profits, but could be a huge step backwards for the world's poor."*—Salil Shetty, Chief Executive at Action Aid

Third World exports

Fostered, in part, by the need to generate foreign income to pay heavy debts to industrialised countries, dependence on export markets for cash crops has undermined local food security. In Kenya, for example, it was reported in 1994 that "the growing of luxury vegetables and flowers had overtaken potatoes, carrots, maize and cabbages for local consumption. On the slopes of Mount Kenya, land had been cleared for growing these luxury crops, thereby depriving local subsistence farmers of water for their crops and causing food prices to rise steeply." [75]

Trade in certain products, however, has existed for hundreds if not thousands of years, and whole communities depend for their livelihoods on these exports. In a move which threatens to undermine these exports, industrialised countries are now using genetic engineering to develop products which used only to be available from the tropics.

- Millions of small-scale cocoa farmers in West Africa, for example, are now vulnerable to the development of genetically-engineered cocoa butter substitutes.[76]

- Vanilla accounts for 10% of Madagascar's export earnings. It is estimated that 70,000 farmers in Madagascar currently face ruin because vanilla is being grown under tissue culture in biotech laboratories.[77]

- Oilseed rape has now been genetically engineered to produce lauric acid, which is traditionally derived from coconut and palm kernel oils, and is used in soap and cosmetics. If the multinationals currently sourcing their lauric acid from coconuts and palm oil were to source it instead from the transgenic oilseed rape, it could affect the livelihoods of millions of people. For example in the Philippines (the world's largest exporter of coconut oil), the coconut industry provides employment for about 30% of the population—some 21 million people.[78]

Terminator technology

'Terminator technology' is the name given to a technique which genetically disables plants to make them infertile. The seeds from which the plants are to be grown are treated with a chemical stimulant such as the antibiotic tetracycline. They are then planted, and a genetic process, triggered by the stimulant, leads to the production of a toxin just before the growing plant's own seeds mature, making them sterile.[79]

Terminator technology is being jointly developed by Delta & Pine Land Co and the United States Department of Agriculture (USDA). USDA spokesperson Willard Phelps freely admits that its primary function is "to increase the value of proprietary seed owned by US seed companies and to open up new markets in second and third world countries"—countries where it is notoriously difficult to stop farmers from saving seeds and enforce patenting regimes. As USDA molecular biologist Melvin J. Oliver, primary inventor of the Terminator, explains: "Our mission is to protect US agriculture and to make us competitive in the face of foreign competition. Without this, there is no way of protecting the patented seed technology." [80]

It has been argued that concerns about terminator technology are irrational, bearing in mind the fact that hybrid crops have been grown for over 50 years. Hybrid crop varieties usually have high yields in the first year and are sterile or lose their vigour in the second generation, motivating farmers to purchase fresh seed each year.[81] The seeds of major crops such as rice, however, are not easily hybridised, and have been saved and planted year after year. This new technology, according to Delta and Pine Land Co, has the prospect of "opening significant worldwide seed markets to the sale of transgenic technology, for crops in which seed currently is saved and used in subsequent plantings."

Camila Montecinos, from the Centro de Educación y Tecnología in Chile, points out that there are key differences between terminator technology and hybridisation. "The theory behind hybridisation is that it allows breeders to make crosses that couldn't be made otherwise and that are supposed to give the plant higher yields and vigour. The results are often disappointing but that's the rationale. In the case of Terminator technology, there's absolutely no agronomic benefit for farmers. The sole purpose is to facilitate monopoly control and the sole beneficiary is agribusiness." [82]

Proponents of the technology argue that farmers will not buy seed that does not bring them benefits. Hope Shand and Pat Mooney from the Rural Advancement Foundation International (RAFI), suggest that choice is not always so simple. "Market choices must be examined in the context of privatisation of plant breeding and rapid consolidation in the global seed industry. The top ten seed corporations control approximately 40% of the commercial seed market. Current trends in seed industry consolidation, coupled with rapid declines in public sector breeding, mean that farmers are increasingly vulnerable and have far fewer options in the marketplace." [83]

The widespread use of proprietary seed has already led to huge losses in genetic diversity. This trend will certainly be exacerbated if farmers are no longer able to save their seeds. Farm-saved seed is central to the livelihoods of 1.4 billion farming households around the world. Concerns about the impact that terminator technology will have on poor farmers have led the United Nations-funded Consultative Group on International Agricultural Research to recommend that its sixteen member institutes ban the technology in their crop improvement research programmes. [84]

In October 1999, after 18 months of intense opposition to terminator technology, Monsanto declared that it had no plans to commercialise terminator seeds, and then announced that it was no longer going to buy Delta and Pine Land (joint holders of the original terminator patent. The USDA was quick to defend the terminator patent, pledged to continue research and said that the government was close to finishing negotiations on a marketing license with Delta & Pine Land for commercialising the technology. Importantly, Monsanto also admitted that it had no plans to abandon closely related research targets that could enable it to switch on (or off) other genetic traits vital to a crop's productivity. [85]

RAFI has discovered that Terminator technology may just be the tip of the iceberg.[86] AstraZeneca, for example, has a patented technology that makes plant growth and germination dependent upon repeated application of proprietary chemicals. "Essentially," says RAFI's Edward Hammond, "they're talking about the manufacture of junkie plants that are physically dependent on a patented chemical cocktail." In February 1999 AstraZeneca categorically stated that it abandoned the development of its seed sterilisation technology in 1992. However RAFI discovered that ExSeed, a joint venture between AstraZeneca and Iowa State University, won a new seed sterilization patent on 11 August 1997, based on a claim made in 1995—three years after AstraZeneca's research was to have been abandoned.

Novartis has a technique which the company describes as "inactivation of endogenous regulation". This patented technique is used to turn off genes which are critical to a plant's ability to fight off infections from many viruses and bacteria. "The only way to turn them back on and fix these 'damaged goods' says Hammond, "is—well, you guessed it—the application of a proprietary chemical."

RAFI also reports that over two dozen patents on genetically sterilised or chemically dependent seeds have recently been obtained by 12 different companies. As RAFI points out, "Companies don't patent for the fun of the paperwork and paying lawyer's fees . . . Unless it is banned by governments, Terminator is going to happen, and probably sooner rather than later." [87]

Chapter Four
Patenting Life

"To justify patenting living organisms, those who seek such patents must argue that life has no vital or sacred property . . . But once this is accomplished, all living material will be reduced to an arrangement of chemicals or mere compositions of matter."—Ted Howard, amicus brief before the US Supreme Court[1]

Thomas Jefferson introduced America's first patent act over two hundred years ago; this allowed inventors to patent "any new and useful art, machine, manufacture or composition of matter, or any new useful improvement [thereof]."[2] In order to patent an object, the inventor has to prove that it has never been made before, involves a 'non-obvious' inventive step, and that it serves some useful purpose. Until recently, these criteria excluded living organisms, which were always regarded as discoveries of nature, and therefore unpatentable.

In 1980, however, this was all set to change. In the landmark case of Diamond v Chakrabarty, the US Supreme Court ruled that a life form (a bacterium genetically engineered to digest oil) could be patented. Chief Justice Warren Burger declared that the "relevant distinction is not between animate and inanimate things but whether living products could be seen as human-made inventions."[3] But can a living organism properly be regarded as a human invention? Even Chakrabarty, the supposed inventor of the bacterium, admitted that he "simply shuffled genes, changing bacteria that already existed—it's like teaching your pet cat a few new tricks." [4] Genetic engineers do not create life; they manipulate genes.

Key Dismukes, from the National Academy of Sciences in the United States, was highly critical of the judgement. "Let us get one thing straight. Ananda Chakrabarty did not create a new form of life; he merely intervened in the normal processes by which strains of bacteria exchange genetic information, to produce a new strain with an altered metabolic pattern. 'His' bacterium lives and reproduces itself under the forces that guide all cellular life . . . We are incalculably far away from being able to create life *de novo*, and for that I am profoundly grateful. The argument that the bacterium is Chakrabarty's handiwork and not nature's wildly exaggerates human power and displays the same hubris and ignorance of biology that have had such a devastating impact on the ecology of our planet." [5]

This extraordinary decision by the US Supreme Court heralded a new era in which living organisms could be patented, and paved the way for the enclosure of the biological commons. Once a shared heritage, the gene pool of plants, animals and humans was now a commodity waiting to be bought and sold.

The significance of the ruling was not lost on corporate investors. A few months later, on 14th October, a recently formed biotech company called Genentech offered a million shares of stock to the market at $35 per share. After just 20 minutes, the shares were being sold at $89; by the end of the day, the company had raised $36 million. Genentech had not yet introduced a single product onto the market. [6] In the words of Jeremy Rifkin, president of the

Foundation on Economic Trends, genes had been identified as the "raw resource for future economic activity". For those with the necessary technology and capital, the race to patent life had begun.

In 1985, the US Patent and Trademark Office (PTO) extended the Chakrabarty ruling to allow for the patenting of genetically engineered plants, seeds and plant tissue. A patent, which usually lasts for 17-20 years, gives the patent holder exclusive rights to exploit an invention for commercial gain. What this means in the case of genetically engineered crops, for example, is that farmers have to pay a license fee and royalties for the use of GE seed and all seed produced from the plants for duration of the patent.

By 1990, 50% of plant patent applications in Europe were coming from just eight multinational corporations, and a third from just three companies: Monsanto, Ciba-Geigy and Lubrizol.[7] Since 1985, these companies have been pushing the boundaries of patent law even further, staking territorial claims to cover entire species of plant and animal, consolidating their dominant position as a means to block research and competition.[8] According to the *Wall Street Journal*, in the United States at least one company has been created whose "main business is buying up broad patents and then suing other companies for alleged infringements."[9]

• In 1994, the company Agracetus was awarded a European patent which covered all genetically engineered soybeans. Rival companies, including Monsanto, were outraged and immediately challenged the patent, saying that it would result in just one company having an effective monopoly over all transgenic soybeans. Monsanto argued that "the alleged invention lacks an inventive step" and was "not . . . novel". In the end the solution for Monsanto was to buy Agracetus, together with the patent, and drop the complaint. As well as the patent on soya, Monsanto now holds a patent in both Europe and the US on all genetically engineered cotton.[10]

- Plant Genetic Systems, a biotech company now owned by AgrEvo, has been granted a patent in the United States for all genetically engineered plants containing the Bt toxin. A patent has been taken out in Europe by the American company Mycogen, which covers the insertion of "any insecticidal gene in any plant".[11]

- A patent has been issued to Sungene in the United States for a variety of sunflower which has a high oleic acid content. Not only does the patent include the genes involved in the oleic acid content, but also to the characteristic itself. Sungene has notified other breeders that the development of any variety of sunflower high in oleic acid will be considered an infringement of the patent.[12]

It is extraordinary that a company can make a single genetic alteration to a plant, and claim private ownership to it as their invention, when the very plants that are being engineered result from thousands of years of careful selection and breeding by farmers around the world.

> "The granting of patents covering all genetically engineered varieties of a species . . . puts in the hands of a single inventor the possibility to control what we grow on our farms and in our gardens. At the stroke of a pen, the research of countless farmers and scientists has potentially been negated in a single, legal act of economic highjack."—Dr. Geoffrey Hawtin, Director General of the International Plant Genetic Resources Institute [13]

Animal patents

In 1987 a Harvard biologist was granted the first patent for an animal. The 'oncomouse' was genetically engineered to predispose it, and all its offspring, to develop cancer, so they can be used for research.[14] The patent on the oncomouse, which is licensed to DuPont (the corporation that financed the research),[15] extends to any other animal genetically engineered to contain genes that cause cancer.[16]

By 1997, over forty animals had been patented, including turkeys, nematodes, mice and rabbits. Hundreds of other patents are currently awaiting approval, including patents on pigs, cows, fish, sheep and monkeys.[17]

- Tracey the sheep has had human genes engineered into her DNA so that she produces a human blood-clotting agent called alpha-1-antitrypsin in her milk. The patent is held by Pharmaceutical Proteins Ltd (PPL). Their spokesperson described sheep like Tracey as "furry little factories walking around in fields". Tracey's success was said to provide "a strong impetus to the further exploitation of transgenic sheep as bioreactors for the production of large amounts of pharmacologically active proteins".[18]

- PPL has also applied for a broad patent covering all cloned mammals. Dolly the sheep was the first mammal to be successfully cloned, in February 1997: a nucleus taken from a cell from the udder of an adult sheep was implanted into an egg which had had its own nucleus removed. This egg was then transferred into another sheep, where it developed into Dolly, who is genetically identical to the sheep from which the udder cells were taken.[19] The Scottish research team who cloned her applied for a broad patent which would give them exclusive patent rights over all cloned animals.[20]

- A team of researchers from a clinic in South Korea recently announced that they had cloned a human cell. They destroyed it in the early stages of cell division, just after "the fourth cell stage, the stage of embryo development when a test tube embryo is placed back in the uterus, where it then further develops into a fetus".[21] Government advisors in the UK have said that cloning of whole humans should be banned but that experiments using 'cell nucleus replacement' with human cells should be allowed for medical research and could start within a year of the appropriate legislation being passed.[22]

Engineering animals

GE animals, including fish, are produced by microinjection. Fertilised eggs are taken from female animals and then injected with foreign genes which will, in some cases, be incorporated into the DNA. These eggs are then returned to the womb of a surrogate mother where they complete their development. Very few will ever reach adulthood and express the new trait in the desired way.[23]

In 1996, over 60,000 transgenic animals were born in UK alone. Most were destined for biomedical research, engineered to be prone to disease and used as research models upon which to test new drugs and therapies.[24]

In May 1997, scientists from Tokyo announced that they had successfully engineered mice with a gene derived from a jellyfish, which produces a fluorescing green protein. Because so few of the foetuses produced by GE actually turn out to have been successfully engineered, it is thought that this fluorescing green protein could be used as a marker to enable scientists to tell which of the foetuses have incorporated the new DNA.[25]

Other animals are being genetically engineered as 'bioreactors' to produce important chemicals in their milk,[26] and for experiments to assess their potential as spare part factories for human transplant operations (xenotransplants). In 1992 a British scientist engineered pig embryos with a human gene, which reduces the likelihood that the organs will be rejected following transplants.[27] Although this means that organ transplants from pigs may now be possible, there are concerns that xenotransplantation could open up a new route for the transmission of animal viruses to human patients.[28]

Attempts are also being made to engineer cattle, sheep, pigs and chickens to increase their growth rates, have lower fat levels, and more tolerance of diseases common to overcrowded and unhygenic factory farms.[29] Also being researched are pigs and poultry that are more docile and better suited to intensive farming conditions, and even featherless chickens that do not need to be plucked.[30]

Pigs engineered with human growth hormone genes at a USDA research station in Beltsville, Maryland, developed severe arthritis, had spinal deformities, and became blind or cross-eyed.[31] Similar experiments with a bovine growth hormone gene resulted in gastric ulcers, liver and kidney damage, bone and joint problems leading to lameness, loss of co-ordination, sensitivity to pneumonia, damaged vision and diabetic conditions.[32]

In sheep, incorporation of human, ovine and bovine growth hormone genes resulted in disrupted joint development and a diabetes-like condition, suppression of appetite and a shortened lifespan.[33]

Human patents

In the United States, a restriction on the patenting of human beings has so far been based on an interpretation of the Thirteenth Amendment of the US Constitution, the anti-slavery amendment, which prohibits ownership of a human being.[34] However, human life forms not covered by this amendment include embryos and foetuses, genetically engineered human tissues, cells and genes.[35].

In 1976 a leukemia patient named John Moore underwent surgery at the University of California to remove his cancerous spleen. Although he signed a pre-operative consent form which said that his spleen would be destroyed after the splenectomy,[36] his doctor cultured some of his cells and found that they produced a special protein. The cells were cultured to produce cancer drugs whose long-term value for the pharmaceutical industry was placed at over $3 billion.[37]

In 1981 Moore's doctor obtained a patent for the cell line and was listed on the application, together with a colleague, as its inventor. John Moore, who later heard that he had been described by the doctor as his "gold mine",[38] filed a lawsuit in 1984, claiming rights to the cells from his spleen. To the relief of the biotech industry, however, in 1990 the California Supreme Court barred the plaintiff, the source of the cells, from obtaining the benefit of the cells' value, but permitted the defendants, "who allegedly obtained the cells from plaintiff by improper means, to retain and exploit the full economic value of their ill-gotten gains free of . . . liability."[39]

"Ultimately, everyone was protected and rewarded: the researcher, the physician, the entrepreneur, even Science," said Moore. "But I knew nothing. What was I? The dehumanisation of having one's cells conveyed to places and for purposes that one does not know of can be very, very painful."[40]

In 1991, the European Patent Office granted a patent on a human gene for the first time, defending its position to the European Parliament by arguing that 'DNA is not life'. The EU then came under intense pressure, in particular from US-based multinationals eager to market and expand their business in Europe, to endorse such patents and harmonise European patent law with countries such as the United States. As a consequence, the 'Directive on the Protection of Biotechnological Inventions' was eventually passed in July 1998. The industry managed to persuade the European Parliament that patents for human genes and cell lines were essential for innovation and progress in the field of medical research. In many cases, however, patents allow companies to profit from research done in the public sector—nearly three-quarters of patents taken out by US corporations in recent years were based on publicly financed research.[41]

Rather than promote medical research for the good of all, patents have the potential to change what was once a field that encouraged the sharing of information and resources into a commercial sector of closely guarded secrets. "I've been at conferences where we have been addressed by patent lawyers and told to stop showing our colleagues our notebooks, or think twice about submitting an abstract at a meeting," says Jonathan King, a genetic researcher at the Massachusetts Institute of Technology. "It's a common experience at scientific meetings for people to withhold information because they have a patent pending. Progress is being slowed down."

A poll of American laboratory directors, published in December 1999, found that a quarter of them had received letters from lawyers acting for biotechnology companies ordering them to stop carrying out clinical tests designed to diagnose Alzheimer's disease, breast cancer and a number of other disorders. Half the laboratories questioned in the survey had been forced to stop the development of screening tests because they knew a patent had been licensed or was pending.

A Massachusetts corporation called Athena Diagnostics, for example, wrote to laboratories informing them of its "exclusive rights to certain tests in the diagnosis of late-onset Alzheimer's disease. These tests are covered under US patent number 5,508,167, a copy of which is enclosed." Athena went on to say that it would be pleased to perform the tests for the published price of $195 per specimen. That is more than twice the price previously being charged by some university medical laboratories, and way beyond the means of some of the researchers operating on government grants, who examine hundreds of samples in the search for new mutations and possible therapies.

Another company, Myriad Genetics, which holds exclusive patents for mutations linked to breast and ovarian cancer (the BRCA 1 and BRCA 2 genes), is reported to have sent letters to a number of laboratories ordering them to stop screening women for these mutations.[42]

The quest for patentable products may also use funds that could better be used for research into preventative health measures. It has been estimated, for example, that at least 90% of human breast cancers are not caused by breast cancer genes, but are triggered by environmental pollutants, diet and lifestyle factors. The belief that genes are the key to understanding and treating disease, however, means that alternative approaches to reducing illness or encouraging health get little or no funding.[43]

• A US-based company called Biocyte holds a patent on all umbilical cord cells from foetuses and newborn babies. The blood cells in the umbilical cord are an important source of stem cells, stem cells being the progenitors of all types of cells in the blood, and are used in therapeutic treatments such as blood and marrow transplantation, and in gene therapies. The patent was awarded simply because Biocyte was able to isolate the blood cells and freeze them. It gives them the right to demand fees from anyone extracting or using umbilical cells from newborn babies, or using any other therapies developed in connection with their use, and also the right to 'refuse access' to the umbilical cells for anyone who is unwilling to pay for the privilege. In June 1999, Biocyte's European patent on the cord cells was revoked after a legal challenge brought by the 'European Campaign On Biotechnology Patents'.[44]

- In 1991, Systemix Inc was awarded a patent giving them rights to all human bone marrow stem cells.[45] The patent not only includes the process used to isolate the stem cells but also covers the stem cells themselves, even though they have not been engineered, manipulated or altered in any way.[46] As Peter Quesenbury, Medical Affairs Vice Chairman of the Leukemia Society of America, commented, "Where do you draw the line? Can you patent a hand?"

Mapping the human genome

The Human Genome Organisation was launched in 1988 as a 15-year, $3 billion international collaborative effort to map the more than 100,000 genes and three billion chemical compounds contained within the human genome.[47] Sequences of DNA are 'tagged' and patented by their 'discoverers', even before there is any understanding about their function. Much as someone might lay claim to a previously unexplored piece of land in case it contains oil or gold, human DNA sequences have been patented in case they turn out to be valuable in the future. It is difficult to estimate just how many human genes have already been patented but, according to one company, the US Patent and Trademark Office has issued more than 1,250 patents on human gene sequences.[48]

• Incyte, a California-based gene sequencing company, has applied for patents on 1.2 million DNA fragments. Another company, Human Genome Sciences Inc, also has patents pending on millions of these fragments. Patent office officials say that there is no legal reason for them not to grant patent rights for sequence tags, and claim that such patents could provide an economic spur to genetic research.[49] Incyte say that their "goal is now to have sequenced, mapped and filed for intellectual property on the novel and most commercially relevant genes by the second half of the year 2000."[50]

• The worldwide market for cell lines and tissue cultures brought in $426.7 million in corporate revenues in 1996.[51] The Human Genome Diversity Project, nicknamed 'the vampire project' by critics, is intent on harvesting valuable DNA samples from from 722 human populations, many of them indigenous peoples.[52] Researchers are collecting blood, tissue and hair samples from, amongst others, the San peoples of the Kalahari, Australian aborigines, the Penan of Sarawak, Latin American Indians, and the Saami of northern Scandinavia.[53]

Biopiracy

Genetic diversity is at its richest in the countries of the Third World, which are estimated to contain over 95% of the world's genetic resources.[54] In what the industry calls 'bioprospecting'—and Third World countries call 'biopiracy'—scouts are sent to these areas to seek out valuable organisms or plants, often drawing upon the wisdom of indigenous peoples. They then take samples back to laboratories where they isolate active ingredients or genetic sequences and patent them as their own inventions.

Traditional knowledge systems, and the people who have cultivated biodiversity over thousands of years, count for less in patent law than routine laboratory procedures. Communities could now end up having to pay multinational corporations for the right to use something that was previously part of their legacy.

- The Neem tree has been used in India for thousands of years, and is valued for its antibacterial and insecticidal properties.[55] Since 1985, US and Japanese corporations have taken out over a dozen patents on these long-appreciated properties of the plant.[56]

- The j'oublie berry from the Gabon in West Africa contains a sweet compound which has been patented by the University of Wisconsin in the hopes that it will make money in the lucrative sweetener market. Despite the fact that the sweet taste of the berries is well known in West Africa, the university claims that the sweet compound (brazzein) is its own invention and admit to no connection with the Gabon.[57]

- According to an article in *Business Week*, when employees of Novo Nordisk go on holiday, they take along soil-collection kits to gather exotic microorganisms. The father of one scientist who worked for the company collected a soil sample from Indonesia which yielded an enzyme that is now widely used by soft-drink suppliers to change starch into sugar. [58] A spokesperson for Monsanto said that the company was recruiting employees "who are traveling somewhere exotic and wouldn't mind digging up a few soil samples for the sake of science" for Monsanto's agricultural screening programs. "You never know what you're going to find or where you're going to find it . . . Nothing's off limits." [59]

Many modern pharmaceuticals are derived from tropical plants. Knowledge about their properties often comes from indigenous communities who have a rich understanding of medicinal plants in their native habitats. The biotech industry argues that this knowledge only becomes valuable once money has been spent on research and commercial products have been developed. This is used to justify the fact that the lands and communities from which these plants originate rarely receive any compensation. The following quote, however, shows that traditional knowledge is worth more to the biotech industry than the above argument suggests:

> *"When we decide to develop a drug, it's already been used in human beings for a long time, in some cases, hundreds of years or more. We have a reasonable assurance that there's less liability as far as safety problems [are concerned] . . . And when you're working on small molecules, that's always a very significant potential problem and an unknown until you get into some pretty expensive animal work or into humans themselves."*
> —G. Kirk Raab, Chairman of Shaman Pharmaceuticals [60]

A report commissioned by Christian Aid estimated that biopiracy was cheating Third World countries out of US

$4.5 billion a year.[61] On the rare occasions that the biotech industry pays for genetic resources, the terms of trade remain unequal and disproportionate.

In 1991, for example, Merck Pharmaceuticals signed a contract in Costa Rica, which is estimated to be home to between 5% and 7% of all the world's species. In exchange for exclusive rights to screen, develop and patent new products from plants, microorganisms and animals in the Costa Rican rainforests, Merck paid $1.1 million towards a local biodiversity programme.[62] With an estimated 500,000 species in Costa Rica this payment works out at about $2 per species—not much for company that had a revenue of $8.6 billion that year.[63] Merck also agreed to give back an unspecified percentage (believed to be 1-3%) of any royalties earned from new products developed from the rainforests. This agreement, which has been touted by the US government and World Bank as a "model" for Third World countries, has many of the qualities that have characterised colonial-style trade over the last 500 years. Some attitudes have changed little since European settlers gave gifts to native Americans in exchange for ownership of the island of Manhattan.[64] At Merck's rate of exchange, the world's genetic resources could be bought for US $20 million.

Patents and the World Trade Organisation

In 1993, half a million Indian farmers in Bangalore protested against plans to implement an international system of intellectual property rights favoured by transnational corporations. This agreement, the Trade-Related Aspects of Intellectual Property Rights (TRIPS), was eventually signed in 1994 and will be administered by the World Trade Organisation (WTO), whose primary agenda is to remove barriers to 'free trade'. Any countries ignoring the statutes of the WTO are liable to be prosecuted, and may be subject to severe punitive action including sanctions or fines.

TRIPS was the brainchild of a coalition of corporations who called themselves the Intellectual Property Committee. It was vigorously opposed by the resource-rich countries of the Third World because it legitimises biopiracy, enshrines it in international law and undermines community rights. "Industry identified a major problem for international trade," says Monsanto's James Enyart. "It crafted a solution, reduced it to a concrete proposal, and sold it to our own and other governments . . . The industries and traders of world commerce have played simultaneously the role of patients, the diagnosticians, and the prescribing physicians." [65]

TRIPS does not require companies to ask for prior consent before accessing biological resources, nor does it demand that patent holders share their benefits with the people or lands from which the genes originate. Under the agreement, countries are obliged to bring their patent laws into line with the industrialised nations by extending them to include living organisms or by setting up equivalent systems of intellectual property rights.

> "What is the principled limit to this beginning extension of the domain of private ownership and dominion over living nature? Is it not clear, if life is a continuum, there are no visible or clear limits, once we admit living species under the principle of ownership? The principle used in Chakrabarty says there is nothing in the nature of a being, no, not even in the human patentor himself, that makes him immune to being patented . . . To be sure, in general it makes sense to allow people to own what they have made, because they artfully made it. But to respect art without respect for life is finally contradictory."—Leon Kass [66]

Chapter Five
Who's in Control?

The 'Life Science' Industry

"In the 20th century, chemical companies made most of their products with non-living systems. In the next century, we will make many of them with living systems."—Jack Krol, Board Chairman of DuPont [1]

As the planetary resources that have sustained the economic growth of the industrial era begin to run out, multinational companies are turning towards genes as the primary raw resource for their future economic activity. In the words of Jeremy Rifkin, these companies now aim to use genetic engineering technologies "to rival the growth curve of the industrial age by producing living material at a tempo far exceeding nature's own time frame and then converting that living material into an economic cornucopia." [2]

This industrial transformation is already well underway, and a new breed of corporations is beginning to emerge. Styling themselves as the 'Life Science' industry, these companies typically hold interests in food, additives, pharmaceuticals, chemicals and seeds.

"The common denominator of our business is biology. The research and technology is applied to discover, develop and sell products that have an effect on biological systems, be they human beings, plants or animals."—Daniel Vasella, CEO of Novartis [3]

Confident that the value of patent monopolies will continue to increase more quickly than the value of physical assets,[4] over the past decade these companies have been investing their considerable resources in a systematic process of acquisitions and mergers. These have concentrated economic power in the hands of just a few of the biggest players, and should be considered in the context of a global system where these companies already wield more power than many nation states: by 1995, of the hundred most powerful 'economies' in the world, 48 of them were multinational corporations, and only 52 of them were countries.[5]

- In December 1998, Germany's Hoechst and France's Rhône-Poulenc merged to form Aventis, "the world's biggest life science company", with combined sales of $20 billion per annum.[6]

- Days later, UK-based Zeneca Group PLC and Astra AB of Sweden announced the largest-ever European merger. At more than $70 billion, the assets of the new company will be larger than the 1997 GNP of 93 countries.[7]

- In March 1999, DuPont announced that it would pay $7.7 billion to buy the remaining 80% stake in Pioneer Hi-Bred International, the world's largest seed company.

- Between 1996–98, Monsanto spent $8 billion on new acquisitions, incorporating seed companies, genetic engineering companies and other related interests.[8] However, faced with huge debts and plummeting stock value, Monsanto was soon forced to find a way to protect its pharmaceuticals business from being adversely affected by the growing opposition to GE foods. In December 1999, it announced that it was to join its pharmaceutical wing with Pharmacia-Upjohn in a $27 billion merger, with

Monsanto's agricultural wing to become a separate entity, with 80% of the stock held by the fused enterprise.

- Other life science companies have acted similarly to protect their pharmaceutical wings. In November 1999, Novartis announced that it was to spin-off its huge agricultural biotech division in a new venture with most of Astra-Zeneca's agrochemical and seeds activities, forming a new company called Syngenta.[9]

The top five biotech companies (Astra-Zeneca, DuPont, Monsanto, Novartis, and Aventis), who account for virtually 100% of the market in transgenic seeds, also account for 60% of the global pesticide market and 23% of the commercial seed market.[10] The acquisition of seed companies is an integral feature of the consolidation underway within the emerging 'life science' industry. It has led to the virtual demise of the independent seed industry in industrialised countries,[11] and near monopolies which help to secure markets for new genetically engineered crops. This, together with sweeping patents and contractual agreements with farmers, grain elevators and processing companies, means that the life science industry has more and more control over the food supply all the way from the laboratory to the dinner plate.

- In November 1998 Cargill, the world's biggest grain exporter, announced a merger that would allow it to control 45% of the global grain trade.[12]

- 40% of US vegetable seeds come from a single source.[13] The top five vegetable seed companies control 75% of the global vegetable seed market.[14]

- By 1999, Delta and Pine Land Co, joint holders with the USDA of the patent on terminator technology, controlled 71% of the US cottonseed market.[15]

"This is not just a consolidation of seed companies, it's really a consolidation of the entire food chain."—Robert T. Fraley, Co-President of Monsanto's agricultural sector [16]

The retail value of global food sales is now estimated at $2,000 billion per annum. When compared to a total of $310 billion for pharmaceuticals, $31 billion for agrochemicals, $23 billion for trade in seeds, and $17 billion for animal health, one can see why the development of genetically engineered food is so attractive to the life science industry. [17]

In the public interest

With few exceptions, governments in industrialised countries have been keen to promote genetically engineered food. Numerous surveys, however, have highlighted a discrepancy between government attitudes and those of the public. People's concerns are frequently dismissed as irrational, and based upon a lack of understanding; yet despite attempts by both government and industry to 'educate' the public, opposition to genetic engineering has continued to grow. [18]

- In June 1998, a MORI poll in the UK revealed that 58% of the people surveyed were opposed to GE food—7% more than in an identical poll two years before. Support for GE had fallen from 31% to 22%. 61% said they did not want to eat GE food, 73% were concerned that GE crops could interbreed with wild plants and cause genetic pollution, and 77% wanted a ban on growing until the impacts of GE crops had been more fully assessed. [19]

- A referendum held in Austria in April 1997 resulted in 1.2 million people, a fifth of the electorate, signing a people's petition to ban genetically engineered foods, deliberate releases of GE organisms and the patenting of life. [20]

- A *Time* magazine poll published in January 1999 found that 81% of American consumers believe genetically engineered food should be labelled. 58% of the people surveyed also said that if genetically engineered foods were labelled they would avoid purchasing them.

"One of the many reasons why people are so concerned," says Dr. Sue Mayer, Director of GeneWatch (an independent UK organisation that monitors developments in the field of genetic engineering), "is that people believe that what they are being asked to accept now is just the thin end of the wedge. People question whether we are really in a position to change things in such fundamental ways—do we really know what we're doing?"

"All the research shows that it is the long-term impacts that most concern people. Some scientists and politicians seem to have the attitude that people's fears are irrational and that they just soak up everything the media tells them. I find this incredibly arrogant because people often have a very sophisticated understanding; far from being irrational, their concerns are rather pragmatic. Experience has taught them that it takes a long time before the effects of new technologies become apparent. They don't trust these overly confident assurances that everything is perfectly safe." [21]

Suzanne Wuerthele, a scientist who specialises in risk assessment, agrees: "It took us 60 years to realize that DDT might have oestrogenic activities and affect humans, but we are now being asked to believe that everything is OK with GE foods because we haven't seen any dead bodies yet." [22]

People are also becoming aware that the future direction of agricultural research is being dictated by commercial interests. In 1998, for example, despite the fact that public demand for organic products is growing so rapidly, the British government spent £54.2 million on research and development into agricultural biotechnology (up to £20.3 million of which was spent specifically on agricultural genetic engineering) compared to a mere £1.8 million on research into organic farming.[23] In the United States, a study by the Organic Farming Research Foundation found that out of 30,000 federally financed research projects, those determined to be organic represent less than 0.1 percent.[24]

- In 1997, organic sales in the United States grew by more than 20% for the 7th year in a row. An industry survey conducted by the biotech company Novartis, showed that 54% of Americans would favour organic products if only they had the choice.

- In the UK, demand for organic products, which has accelerated since the debate about genetic engineering hit the headlines,[25] is growing so quickly that 75% of the organic produce sold has to be imported.

- In France too, according to the Minister of Economy and Finances, farmers can hardly keep up with a consumer demand that rose by 25% in 1996 alone.

- In Denmark, organic foods already claim 15% of the market, and demand is predicted to reach 20% by the year 2000.[26]

The implications of such sustained growth is not lost on the biotech industry, who have been quick to adopt the image of sustainability. In a 1990 article entitled 'Planetary Patriotism', two top executives from Monsanto maintained that caring for the environment while meeting a growing demand for food "requires sustainable agriculture" and that "sustainable agriculture is possible only with biotechnology and imaginative chemistry."[27] Val Giddings, Vice President for Food and Agriculture of the Biotechnology Industry Organisation, says that it is only a matter of time before the industry manages to "penetrate into the organic market."[28]

Organic farmers, however, remain some of the most outspoken opponents of genetic engineering, and have consistently rejected any attempt to dilute the 'GE-free' status of organic standards. This issue was highlighted in spring 1998, when the US Department of Agriculture put forward legislation proposing that genetically engineered food could be labelled as 'organic'. In spite of heavy lobbying by the biotech industry, the USDA was forced to drop its plans after receiving an unprecedented 275,000 letters of complaint.[29]

This growing lack of public acceptance has been identified as the biggest problem now facing the biotech industry.[30] EuropaBio, an association of the leading multinationals involved, responded by launching a multi-million dollar campaign to persuade Europe of the benefits of genetic engineering, and sought the advice of Burson Marsteller, past masters at crisis management. Previous clients of Burson Marsteller included Exxon after the Exxon Valdez oil spill, Union Carbide after the explosion of their chemical plant in Bhopal which killed 15,000 people, Babcock and Wilcox after the failure of the Three Mile Island nuclear reactor, the Indonesian government after ex-President Suharto's army carried out the Dili massacre in East Timor, and other ruling military regimes in Argentina and South Korea.[31]

Burson Marsteller advised EuropaBio that "public issues of environmental and human health risk are communications killing fields for bioindustries in Europe . . . all the research evidence confirms that the perception of the profit motive fatally undermines industry's credibility on these questions." They suggested that EuropaBio should focus instead on "symbols eliciting hope, satisfaction, caring and self esteem".

In a demonstration of considerable faith in the proactive attitude of national governments, Burson Marsteller also suggested that EuropaBio should refrain from partaking in any public debate, leaving it to "those charged with public trust, politicians and regulators, to assure the public that biotech products are safe." [32]

The European Commission, for one, has been most obliging. It set up the European Federation of Biotechnology Task Group on Public Perceptions on Biotechnology, which is specifically designed to deal with public resistance. Millions of pounds of taxpayers' money have been allocated to projects designed to persuade people of the benefits of genetic engineering. The 'FACTT' project, for example, has been granted UK £1 million (with a similar amount being contributed by the industry) to promote the sales of genetically engineered oilseed rape. It aims to bring about "the creation of familiarity with and acceptance of transgenic crops for farmers, extension organisations, processing industry, regulatory organisations, consumer groups and public interest groups." [33]

The blurring of the boundaries between public money and corporate promotion have come about, in part, from the erosion of public funding for independent science. As a direct result of government policies which encourage private enterprise within the public sector, most university research laboratories are now inexorably linked to companies that support them through grant or other funding. For a scientist working in one of these laboratories, a professed ability to attract private investment is often at least as important as research skills or academic qualifications. This dependence on funding from the private sector means that research that fails to attract corporate sponsorship is unlikely to be continued. This significant shift in emphasis within universities and research centres is then reflected in the advice that is given by scientists to national governments.

In the UK, for example, the Agriculture and Food Research Council was replaced by the Biotechnology and Biological Science Research Council (BBSRC), for which the future development and expansion of genetic engineering is a key responsibility. The Executive Director of the BBSRC was also, until May 1999, none other than the CEO of Zeneca, while many other representatives of large corporations sit on its Research and Strategy Boards,[34] where they are able to have considerable influence over the direction of agricultural research in the UK. Public interest groups, on the other hand, rarely have any such opportunity.[35]

Furthermore the BBSRC staff code, which applies to publicly funded research scientists in the UK, contains a clause which effectively prevents scientists from speaking out on controversial issues related to genetic engineering. The clause states: "As the [place of work] is supported by public funds and in view of the nature of its work, there are certain restrictions on employees wishing to engage in political activities. Staff should not become involved in political controversy in matters affecting research in biotechnology and biotechnological sciences."[36]

The close relationships between multinational corporations and national governments are increasingly becoming causes for concern. The United States government in particular has been criticised because many of the people now sitting on key regulatory bodies such as the Food and Drug Administration have strong links to the very corporations that they are supposed to regulate.

A report put out by the Edmonds Institute and the Third World Network contained the following details of

Follow the money

A piece of research published in the *New England Journal of Medicine* in January 1998 highlights the difference in opinion between scientific and medical experts who take corporate money, and those who do not.[37]

Researchers in Toronto, Canada, examined the views of different scientists after the National Heart, Lung and Blood Institute warned doctors that use of a calcium channel-blocker, used to treat high blood pressure and heart disease, increased the risk of deaths from heart attacks.[38] In the ensuing medical controversy, other channel-blockers came under suspicion.

The researchers examined 70 articles written about the controversy, and classified the authors of these articles into three categories: supporters, neutral, and critical. They then mailed surveys to the authors, in which they asked them about their financial ties to drug corporations. When the surveys were returned, the researchers found that 96% of the authors who had written in support of the drugs had financial relationships with manufacturers of calcium-channel blockers, as compared with 60% of the neutral authors, and 37% of the critical authors. They also found that 100% of the supportive authors had financial ties to at least one pharmaceutical manufacturer, compared with 67% of the neutral authors, and 43% of the critical authors.

This study, the first peer-reviewed research of its kind, clearly suggests that people whose careers are tied to the development of a technology should not be expected to hand out impartial advice. "In other words" says Peter Montague, editor of *Rachel's Environment & Health Weekly*, "if you want to understand 'objectivity' in the science and medicine of environment and health these days, the same advice applies as it does in politics: follow the money." [39]

'revolving doors' between the US government and the biotech industry:

—David W. Beier, former head of Government Affairs for Genentech, now chief domestic policy advisor to Vice President Al Gore.

—Linda J. Fisher, former Assistant Administrator of the US Environmental Protection Agency's Office of Pollution Prevention, Pesticides, and Toxic Substances, now Vice President of Government and Public Affairs for Monsanto.

—L. Val Giddings, former biotechnology regulator and (biosafety) negotiator at the US Department of Agriculture (USDA/APHIS), now Vice President for Food & Agriculture of the Biotechnology Industry Organisation (BIO).

—Marcia Hale, former assistant to the US President and director for intergovernmental affairs, now Director of International Government Affairs for Monsanto.

—Michael (Mickey) Kantor, former Secretary of the US Department of Commerce and former Trade Representative of the US, now on Monsanto's Board of Directors.

—Josh King, former director of production for White House events, now director of global communication in the Washington, DC office of Monsanto.

—Terry Medley, former administrator of the Animal and Plant Health Inspection Service (APHIS) of the USDA, former chair and vice-chair of the USDA Biotechnology Council, former member of the FDA food advisory committee, and now Director of Regulatory and External Affairs of DuPont Corporation's Agricultural Enterprise.

—Margaret Miller, former chemical laboratory supervisor for

Monsanto, now Deputy Director of Human Food Safety and Consultative Services, New Animal Drug Evaluation Office, Center for Veterinary Medicine in the FDA.

—William D. Ruckelshaus, former chief administrator of the US EPA, now (and for the past 12 years) a member of the Board of Directors of Monsanto.

—Michael Taylor, former legal advisor to the FDA's Bureau of Medical Devices and Bureau of Foods, later executive assistant to the Commissioner of the FDA, still later a partner at the law firm of King & Spaulding where he supervised a nine-lawyer group whose clients included Monsanto Agricultural Company, still later Deputy Commissioner for Policy at the United States Food and Drug Administration, and now again with the law firm of King & Spaulding.

—Lidia Watrud, former microbial biotechnology researcher at Monsanto in St. Louis, Missouri, now with the EPA Environmental Effects Laboratory, Western Ecology Division.

—Jack Watson, former chief of staff to President Carter, now a staff lawyer with Monsanto in Washington, DC.

—Clayton K. Yeutter, former Secretary of the USDA, former US Trade Representative, now a member of the Board of Directors of Mycogen Corporation, whose majority owner is Dow AgroSciences.

Despite this movement of high level personnel back and forth between industry and government, US Agriculture Secretary Dan Glickman assures us that the United States "will continue to insist on an arms-length, objective testing process that is independent from industry," and that "test after rigorous scientific test" has proven the safety of genetically engineered products.[40] In reality, the US Congress made a strategic decision in the 1980s to deregulate the products of agricultural biotechnology, reasoning that current food laws seemed broad enough to cover them and anxious that "regulatory excess" would "suffocate" the young industry.[41]

In May 1992, before the first transgenic foods came onto the market, the FDA determined that most of the foods produced by genetic engineering should be regarded and regulated as if they were foods produced by traditional methods. This means that, except in certain cases, such as when there are major changes in nutrient composition or incorporation of specific proteins known to cause allergic reactions, genetically engineered foods in the US do not require a pre-market approval process, public notification, or labelling. The industry decides when and whether to consult with the FDA, and it is they who conduct safety tests for their own products, notifying the FDA only if they suspect a problem. Thus it is the very companies who stand to profit who decide whether or not these products are hazardous.[42]

In Europe too, there is clearly a desire to create a regulatory climate which is attractive to the industry. It is estimated that the potential market for biotech related products will reach US $278 billion by the year 2000, with up to 70% of new growth in this field coming from the agriculture and food sector.[43] As Stephen Byers, UK Secretary of State for Trade and Industry, said in his speech at the Biotechnology Industry Association Gala Dinner in January 1999, "The biotechnology industry is exactly the kind of industry we want to encourage in this country. Poverty of ambition has too often held us back . . . Our current annual spend of some £600 million a year on biotechnology research and development speaks for itself."[44]

In keeping with this favourable attitude, the European

Commission has decided that a country may only reject an approval for a genetically engineered organism if they are able to provide strong "scientific evidence" of harm. Once an approval has been granted in one country, the rules then oblige all other countries in the EU to accept that decision. In practice this means that the countries least concerned about potential hazards could set the EU-wide norms for approval.

This democratic deficiency within the European regulatory system was highlighted in the spring of 1997 when the European Commission decided to approve a variety of transgenic maize produced by Novartis, even though 13 out of 15 member states had voted against it.[45] On 8th April, the European parliament voted resoundingly (407 in favour, 2 against) for a resolution condemning the Commission for "a lack of responsibility" in approving the maize in spite of the fact that "serious doubts remain as to [its] safety." The resolution added that "trade considerations have obviously dominated the decision-making process so far."[46]

The precedence given to trade was demonstrated in February 1999, when a small group of grain-producing countries, led by the US, blocked an international treaty called the Biosafety Protocol, which had been designed to regulate the trade and safety assessment of genetically engineered organisms. The US refused to accept the treaty because it argued that Third World attempts to introduce strict safety assessments and socio-economic considerations were endangering free trade.

Speaking after the negotiations collapsed, the Ethiopian official leading the African Delegation, Dr. Tewolde Berhan Gebre Egziabher, said that African countries were "absolutely united" in resisting US plans to "decide what we eat".[47] When asked why the United Kingdom, amongst European countries, had been one of the least supportive of the needs of Third World countries, he remarked that "the position of the UK delegation is shaped by corporate interest, probably reinforced by transatlantic pressure."

"All technologies involve risk", he added. "Since genetic engineering manipulates the basis of life, the risks involved are more frightening than any other developed so far. It is therefore essential for those of us who are the poorest of the world, and thus most vulnerable, to require a regime which assigns liability and ensures redress. We feel it is unjust of the richest of the world to expect us to bear the risks of their experimentation." [48]

Although there is broad agreement between governments in the United States and Europe about the importance of such trade considerations, and a desire on both sides of the Atlantic to encourage the biotech industry, there have been serious disputes over certain aspects of policy. Firstly, there is the fact that massive resistance from European citizens has forced their governments to be more cautious. This has led to delays in the approval of new transgenic crops, and even outright bans in countries such as Austria and Luxembourg, which have been forcefully challenged by the United States. Secondly, there is the issue of labelling and segregation.

Labelling

When people began to realise they were eating genetically engineered food without their knowledge or consent, there were immediate calls for segregation and labelling. Survey after survey showed that the vast majority of people wanted comprehensive labelling of GE food, even if they did not mind eating it. It was also argued that labelling would be essential in order to be able to trace any health problems that may arise.

In May 1998, however, Codex Alimentarius, the UN body responsible for establishing international rules on food policy, rejected these demands in favour of a much more limited labelling regime that suited the food and genetic engineering industries.[49] As Julian Edwards, Director General of Consumers International, which represents 245 consumer organisations in 110 countries, pointed out: "One of the ironies of this issue is the contrast between the enthusiasm of food producers to claim that their biologically engineered products are different and unique when they seek to patent them and their similar enthusiasm for claiming that they are just the same as other foods when asked to label them."[50]

The concept of 'substantial equivalence' was used to argue that genetically engineered food was 'equivalent' to food produced by any other means, and that labelling would therefore be discriminatory and constitute an illegal trade barrier. Biotech companies were also concerned that segregation would need to be introduced in order to implement labelling schemes. This would raise the cost of genetically engineered ingredients, potentially making them uneconomical for the food industry.

However in 1998 the EU, having been under sustained pressure from the public, introduced a partial labelling scheme which covered transgenic soya and maize. Most processed food in Europe contains genetically engineered ingredients from soya and maize, the majority of which are derivatives, such as soya oil, lecithin and corn (maize) syrup. Yet these derivatives were excluded from the new labelling scheme, because the industry argued that most of the genetically engineered DNA would be destroyed when food is processed. Surveys have found that even so, most people want the right to know if the method of production used for food they are eating involves genetic engineering, and they may have ethical reasons or environmental concerns that make them want to avoid it.

Despite the fact that this labelling regime was widely criticised in Europe and regarded as inadequate, the United States government was adamant that there should be no labelling or segregation whatsoever. US Trade Representative Charlene Barshevsky estimated that the EU proposal for segregating and labelling genetically engineered food could disrupt $4-5 billion in annual US agricultural exports.[51] In June 1997, she gave a stern warning to European leaders that they could expect "at the minimum" punitive action through the World Trade Organisation if they allowed domestic concerns over biotechnology to disrupt US agricultural trade.

"We will not tolerate segregation," said US Agriculture Secretary Dan Glickman. "We will not be pushed into allowing political science to govern these decisions. The stakes for the world are simply too high[52]. . . We will lead the fight against those who represent what I believe is a know-nothing position on these issues . . . we will not allow passion to trump reason on this issue." [53]

There is evidence that the United States government has been applying pressure on other countries to reject labelling regulations. The Japanese government, for instance, has been coming under increasing pressure to introduce labelling, having received several million signatures from the public. In September 1998, Charlene Barshevsky was sent to Tokyo to discuss the issue with Shoichi Nakagawa, the Japanese Agriculture Minister. She warned him that that any plans to label genetically engineered food were unacceptable and could jeopardise trade relations between Japan and the United States.[54] Similarly, a New Zealand cabinet document from February 1998 showed that the US had threatened to pull out of a potential free-trade agreement with the New Zealand government because of its plans to test and label transgenic food. The document stated: "The United States have told us that such an approach could impact negatively on the bilateral trade relationship and potentially end any chance of a New Zealand-United States Free Trade Agreement."[55]

By the end of 1999, however, mounting public pressure had forced regulatory authorities in Australia, New Zealand, South Korea, and Japan to begin to implement mandatory labelling of GE foods, and similar demands for labelling were building in the US, Malaysia and the Philippines.[56]

Products approved by 1998

By December 1998, the following genetically engineered products had received approval in the US: herbicide-resistant canola (oilseed rape), radicchio, maize, cotton, and soybeans; insect-resistant maize, cotton and potatoes; virus-resistant papaya, potato and squash; canola (oilseed rape) designed to produce high concentrations of lauric acid; tomatoes engineered to delay their ripening, or have thicker skins; a rabies vaccine; a bacterium designed to enhance nitrogen fixation in the soil, and a genetically engineered growth hormone (rBST/rBGH) designed to boost milk production in dairy cows.[57]

Eighteen genetically engineered products had been granted marketing approval in the EU by December 1998. Of these, the only ones to have received unanimous approval by all the member states were two varieties of genetically engineered carnation: one with 'improved vase life', and one with altered colouring. All the other approvals have been disputed—products officially approved as 'safe' have subsequently been banned in certain countries, while the introduction of many of the food crops which have been approved are now subject to delays as a result of concern about their impacts on health and the environment.

Besides the carnations mentioned above, these eighteen products include herbicide-resistant tobacco, maize, chicory, soybeans, and oilseed rape; insect-resistant maize; and pig and rabies vaccines.[58]

Genetically engineered ingredients already in European shops include soybeans, maize and tomatoes, riboflavin (vitamin B2) and a yeast, as well as a variety of other additives and at least a dozen enzymes produced from genetically engineered microorganisms, such as chymosin, a vegetarian rennet now used in most cheese. GE enzymes are not covered by the labelling or regulatory requirements that apply to other genetically engineered food, and are used widely by the processing industry in foods as diverse as fish, egg and meat products, beverages, biscuits, cakes and bread.[59]

We know what's best for you

When she was asked whether she felt that people should be given the choice of eating GE food or not, Janet Bainbridge (chair of the UK Advisory Committee on Novel Foods and Processes) replied that they should not because "most people don't even know what a gene is." She then went on to say, "sometimes my young son wants to cross the road when it's dangerous. Sometimes you just have to tell people what's best for them."[60]

What's in the pipeline?

"Within five years—and certainly within ten—some 90-95% of plant-derived food material in the United States will come from genetically engineered techniques."—Val Giddings, Vice President for Food and Agriculture of the Biotechnology Industry Organisation [61]

Most of the genetically engineered crops already on the market have been designed to be resistant to herbicides or insects. Over the next few years, the industry plans to introduce more crops with 'quality traits' perceived as benefits for consumers or the food processing industry. An example of this is the attempt to engineer fruit and vegetables that ripen more slowly, allowing them to be transported over greater distances and kept for longer on supermarket shelves without losing the appearance of being fresh. [62]

Other kinds of food on their way include the so-called 'functional foods' and 'nutraceuticals', which claim to enhance health and wellbeing. Examples include foods with added vitamins and altered nutritional values or 'low-fat crisps' from potatoes that have a higher starch content and less water, so can be fried in less oil. [63]

According to industry analysts, corporations such as DuPont, Kellogg, ConAgra, Mars, and Astra-Zeneca could soon be making $29 billion a year from foods which purport to have healthy properties. [64] These companies will be allowed to make claims about the 'health-giving' value of these foods without actually having to demonstrate their clinical efficacy. [65] A quick-fix attitude to health prevalent in modern times could be of considerable benefit to the industry, as they aim to use these 'nutraceuticals' to attract consumers back to the genetically engineered foods they have so far rejected. [66]

Chapter Six

A Case Study: Milk and GE Growth Hormones

In autumn 1996, award-winning reporters Steve Wilson and Jane Akre were hired by a Florida TV station to make a series on a genetically engineered hormone called rBST (also called rBGH). Produced by Monsanto, rBST is injected into nearly a third of all dairy cows in the USA to boost milk production.[1]

After more than a year's work, and just three days before the series was scheduled to go on air, Roger Ailes, head of Fox News, received the first of two letters from lawyers representing Monsanto, claiming that Monsanto would suffer "enormous damage" if the series ran. It urged Ailes to involve himself directly in an effort to "get the facts straight" about rBST and added "there is a lot at stake in what is going on in Florida, not only for Monsanto, but also for Fox News and its owner."

Fox News cancelled at the last minute, despite having run a major advertising campaign to promote the series.[2] A second letter followed, warning of "dire consequences" for Fox TV if the series was broadcast in its present form.

There followed nine months of postponements, bitter arguments and 73 rewrites.[3] "We were repeatedly ordered to go forward and broadcast demonstrably inaccurate and dishonest versions of the story," said Wilson. "We were given those instructions after some very high-level corporate lobbying by Monsanto and also, we believe, by members of Florida's dairy and grocery industries.[4] Nowhere in any of the dozens and dozens of versions we've written did any Fox manager or lawyer ever point to even one error of fact." [5]

Akre and Wilson say that their new general manager, David Boylan, told them "We paid $3 billion for these television stations. We will decide what the news is. The news is what we tell you it is." When the two said that they would file a formal complaint, they were offered a separation agreement binding them to silence, together with very large cash settlements if they promised to go away and keep quiet. They refused and were fired.

Four months later they filed their own lawsuit against the TV station for firing them for refusing to broadcast false reports,[6] and for demanding that they include known falsehoods in their series on rBST.[7]

"Solely as a matter of conscience", said Akre, "we will not aid and abet their effort to cover this up any longer. Every parent and every consumer has the right to know what they're pouring on their children's morning cereal.[8] We have every confidence that a jury will agree. And when it does, after we're reimbursed for our lost salaries and legal fees and other costs, every nickel over and above that will be donated to a journalism organisation that can support the next journalist who has to choose between his job and telling the truth," says Jane Akre.[9]

Wilson adds, "We need to know—and more importantly viewers need to know—to what extent news broadcasts are shaped and information deliberately withheld, when news might reflect badly on an advertiser or potential litigant as big as Monsanto."[10]

So what was it that Monsanto didn't want people to know? The US Food and Drug Administration (FDA), who gave Monsanto permission to market rBST in 1993, insist that there is nothing to worry about. They claim that "no significant difference has been shown between milk from rBST-treated and non-rBST-treated cows".[11] They even threatened dairies who didn't use the drug with legal action if they put 'rBST-free' labels on their milk, arguing that as there was 'no difference' between the two types of milk, such a statement would constitute 'false labelling'.[12]

The FDA official responsible for the agency's labelling policy, Michael R. Taylor, has been described as 'a classic product of the revolving door'.[13] After four years as Executive Assistant to the Commissioner of the FDA, Taylor became partner in a Washington-based law firm called King & Spaulding, which then represented Monsanto while it was seeking FDA approval for rBST. Taylor returned to the FDA in 1991 to become assistant commissioner for policy, and in February 1994 he signed a Federal Register notice warning grocery stores not to label milk as 'rBST-free'.[14]

Days later King & Spaulding filed a lawsuit on behalf of Monsanto against an Iowa Dairy cooperative and a milk and icecream company in Texas for labelling their milk in this way.[15] Monsanto then mailed letters to over 2000 retailers, saying that they had "taken action against one milk co-operative/processor for advertising and promotional activities about bovine somatotrophin [rBST] that we believe are misleading." They also repeated this message in a 30-page legal memorandum sent to 4,000 food processors and dairy cooperatives.[16]

In addition, according to the US-based campaign 'BioDemocracy', Monsanto "threatened school boards with lawsuits if they ban rBST [milk] from school cafeterias, lobbied against rBST labelling bills in Congress and states, and threatened states with lawsuits if they passed rBST labelling laws."[17]

Monsanto and the FDA have repeatedly asserted that the basis for their stand against labelling was that "there is no evidence that hormonal content of milk from rBST treated cows is in any way different from cows not so treated." [18] There is in fact considerable evidence showing that residues of the genetically engineered hormone are left in the milk of treated cows. [19] Once injected into the cows, BST stimulates the production of an 'Insulin-like Growth Factor' (IGF-1), a hormone already present in the blood, which causes cells to divide. [20]

As Monsanto themselves admitted in the scientific journal *The Lancet* in 1993, IGF-1 can be present at up to five times normal levels in milk from treated cows. [21] This means that if you consume products made from rBST milk, you are likely to end up with higher levels of IGF-1 in your blood too, because in the presence of casein (the principal protein in cows' milk), IGF-1 resists digestion and is well absorbed across the intestinal wall. [22]

But is this significant? What are the effects of increased levels of IGF-1 on human health? IGF-1 is associated with larger relative risks for common cancers than any other factor yet discovered. [23] Research has shown that women with relatively small increases in their blood-levels of IGF-1 are up to seven times more likely to develop post-menopausal breast cancer than women with lower levels, [24] and high levels of IGF-1 in the blood are the strongest known risk factors for prostate and colon cancer. [25]

Yet Monsanto still maintain that rBST is perfectly safe and "is the single most-tested product in history." [26] Canadian government scientists, however, tell a different story: "The usually-required long-term toxicology studies to ascertain human safety were not conducted. Hence, such possibilities and potential as sterility, infertility, birth defects, cancer, and immunological derangements were not addressed." [27]

Professor Samuel Epstein from the University of Illinois is not impressed: "With the complicity of the FDA, the entire nation is currently being subjected to an experiment involving an age-old dietary staple . . . it poses major potential health risks for the entire US population."[28]

This view was shared by Dr. Richard Burroughs, staff veterinarian and senior scientist at the FDA, who was fired in 1989 when he expressed concerns about the safety of the hormone. "I was told that I was slowing down the approval process. It used to be we had a review process at the FDA. Now we have an approval process. I don't think the FDA is doing good, honest reviews. They've become an extension of the drug industry." [29]

After Burroughs was fired, the official FDA position on rBST was crafted by Dr. Margaret Miller, a former Monsanto researcher who was reportedly still publishing papers on the hormone with Monsanto while she was working for the FDA.[30]

By 1990, the FDA had published in *Science* magazine [31] a justification for its conclusion that milk from cows treated with rBST was "safe for human consumption".[32] They referred to a 90-day feeding trial carried out by Monsanto, and claimed that the rats in the trial showed "no toxico- logically significant changes".[33]

Canadian government scientists, however, state that the FDA misreported the results of the feeding trial, and say that in fact 20% to 30% of the rats fed the hormone in high doses developed primary antibody responses to it.[34] Such a response shows that the immune system has detected and responded to an alien substance entering the body, indicat- ing that the rBST was absorbed into the rats' bloodstreams. Male rats in the study also developed cysts on their thyroids.

So how was it that the FDA reported that there were "no . . . clinical findings" in the Monsanto rat study?[35] According to John Scheid from the FDA's Centre for Veterinary Medicine, the FDA never actually examined the raw data from the feeding study.[36] Instead they based their safety assessment on Monsanto's summary of the feeding trials, thereby breaking their own regulations.[37] After an investigation by the US Government Accounting Office, Vermont Congressman Bernie Sanders said that the findings proved "the FDA allowed corporate influence to run rampant in its approval" of the drug.[38]

Controversy has not been limited to the FDA. Senator Mira Spivak, whose committee investigated the approval process of rBST in Canada says that Canadian health officials provided her staff with a copy of the Monsanto feeding study in which the information about the potentially troubling effects of rBST on rats had been "blocked out".[39]

Furthermore, six scientists at the Health Protection Branch, responsible for drug approvals in Canada, alleged that they were pressed to approve rBST by their supervisor despite their fears that it wasn't safe. One scientist testified that the newly appointed Director threatened to transfer them to another government department, where "they would never be heard of again" if they did not speed up the approvals.[40]

The Senate committee was also told that Monsanto offered government scientists a bribe of $1 to $2 million on the condition that Monsanto "receive approval to market their drug in Canada without being required to submit data from any further studies or trials."[41] Monsanto claimed that the bribery allegation was "a blatant untruth", and said that Canadian regulators just didn't understand that the offer of the money was for research.[42]

The Canadian scientists' report was subsequently sent for a "completely objective and arm's length review" by a panel of experts, one of whom, Rejeanne Gougeon, served as a consultant to Monsanto from 1993 until 1998, and published a paper in 1994 recommending that the

Canadian government approve rBST.[43] In January 1999, however, after eight years of deliberation, the Canadian government finally refused marketing approval for rBST because of serious concerns about animal welfare.

A panel of animal health experts appointed by the Canadian Veterinary Medicine Association found that cows injected with the hormone have a 18% higher rate of infertility, a 50% increase in lameness, are 25% more likely to suffer from mastitis (udder infections), and have reduced life expectancy.[44]

In March 1999 EU scientists announced that they had come to the same conclusion: "BST administration causes substantially and very significantly poorer welfare because of increased foot disorders, mastitis, reproductive disorders and other production related diseases. These are problems which would not occur if BST were not used and often result in unnecessary pain, suffering and distress. From the point of view of animal welfare, including health, the Scientific Committee on Animal Health and Animal Welfare is of the opinion that BST should not be used in dairy cows."[45]

More mastitis means more antibiotics, and more antibiotics mean higher residues of antibiotics in milk from rBST-treated cows. This is turn may encourage antibiotic resistance in bacteria, a serious health risk to both cows and humans[46]—especially with emerging bacterial strains that are now resistant to all known antibiotics.[47]

According to the FDA, before a drug can be approved for use in animals "the company must show that the drug is effective and safe for the animal."[48] However, according to the report by the Canadian government, "evidence from the animal safety reviews were not taken into consideration."[49]

The EU scientists agree: "Questions about the effects of elevated IGF-1 levels in the cow on the welfare of the cow, or the welfare of the calf *in utero*, appear not to have been investigated. Neither have questions about the effects of elevated IGF-1 levels in milk on the welfare of calves which drink the milk." [50]

The Canadian government scientists also say that "there are reports on file that Monsanto pursued aggressive marketing tactics, compensated farmers whose veterinary bills escalated due to increased side effects associated with the use of rBST, and covered up negative trial results." [51]

Such was the experience of three British scientists who analysed trial results from cows treated with rBST. They claim that Monsanto blocked publication of their 1991 paper on the hormone's links to increases in white blood cell counts, which result from infection and inflammation. [52]

Erik Millstone, Eric Brunner and Ian White analysed the data, and found much higher levels of white blood cells and pus in rBST milk than were reported by Monsanto from the same data. "I have the distinct impression that they didn't want our analysis published because it raises questions about the safety and desirability of their product, which they seem not to want raised," said Millstone. [53]

BST increases milk production by about 10–15%. [54] However, overproduction of milk has been a major problem for the US dairy industry since the 1950s: the annual subsidy per dairy cow in the United States now exceeds the per capita income of half the world's population. [55] Paradoxically, while Monsanto receives an estimated $300-500 million each year in the United States from a hormone that boosts milk production, taxpayers in the US spent an average of $2.1 billion each year between 1980 and 1985 buying up surplus milk to prevent its price from plummeting. [56]

Although it is understood that imports of dairy products from the US are likely to contain rBST-treated milk, the EU currently prohibits the sale of rBST to farmers in Europe. The United States, however, knew that this moratorium could be challenged if Codex Alimentarius would declare that BST was safe, Codex being the UN body which sets international guidelines on food policy in relation to trade. The standards set by Codex are those used by the World Trade Organisation, so if Codex were to conclude that the hormone was safe, the US government would then be able to use the powers of the WTO to force countries such as Canada and the members of the European Union to accept it.

Codex met to discuss the approval of the hormone in June 1999. By then, however, the controversy over the safety of rBST had grown to such an extent that the governments represented at the meeting were unable to come to a consensus on whether to adopt a draft standard for 'maximum residue limits' for BST in milk. In the end they were forced to delay the decision until a consensus could be reached.[57]

One of the committees advising Codex on the safety of rBST is the Joint Expert Committee on Food Additives (JECFA). According to a report in *The Observer*, Dr John Hermann, a former employee of the FDA who heads the secretariat of the JECFA, admitted that confidential documents had been leaked to Monsanto by Dr Nick Weber, an FDA official who sits on the panel. Consumers' International have demanded that the JECFA withdraw their approval of rBST claiming that Monsanto's privileged access to restricted documents has 'damaged the objectivity and credibility' of the investigation of the hormone.

There are further concerns about the impartiality of the JECFA because the scientist who drafted the first report on rBST for the committee was none other than Margaret Miller,[58] the former Monsanto employee who drew up the position statements on rBST for the FDA.[59]

Canadian Senator Eugene Whelan echoed the sentiments of many when he said that he would not be impressed by any testimony quoting the WHO, FAO or Codex because "the big companies sit behind them and tell them what to do".[60] Of the 111 organisations that can send observers to Codex meetings, 104 of them are groups funded by the industry, six are health and nutrition foundations and just one is a consumer organisation.[61]

British MP Alan Simpson was even more explicit when he spoke out in parliament in March 1999: "The history of Monsanto's interests in bovine somatotropin milk and genetically modified crops is littered with the company buying its way into public policy decisions in its favour . . . Monsanto faces a ruling either from us or from the European Union collectively that says that we are deeply unhappy about removing the ban on BST milk, because of the damage to livestock and the potential damage to human beings."

He continued: "The company has manipulated the rules of the vetting agencies in a way that is technically brilliant, but ethically corrupt and degrading for humanity. We have to be prepared to make a stand against that . . . As we debate the relationship between Britain, Europe, the USA and the World Trade Organisation, we must understand that people have given up waiting for a parliamentary lead . . . If the United States of America and the food corporations are to threaten us with a trade war and World Trade Organisation rulings that define our actions as illegal, we must reply that the British public would probably deem a trade war to feed safe food to ourselves and our children a war worth fighting. Rulings may go in their favour, but if Monsanto and the USA win the right to dump unsafe foods in United Kingdom markets, we can overrule that right with a civic right to dump those products in the sea."[62]

Chapter Seven
Turning the Tide

"If you think you are too small to make a difference, try sleeping in a closed room with a mosquito"—African Proverb

In June 1998, Monsanto launched a £1 million advertising campaign designed to "encourage a positive understanding of biotechnology". Months later, a report written for Monsanto by Stanley Greenberg, polling adviser to Bill Clinton, Tony Blair and Gerhard Schroeder, revealed that the advertising campaign "was, for the most part, overwhelmed by the society-wide collapse of support for genetic engineering in foods." There were, said the report, "large forces at work that are making public acceptance problematic."[1]

These "large forces" consist of concerned individuals, local initiatives, church groups, conservation bodies, scientists, farmers, consumer organisations, environmentalists, and others. The annual revenue of the biggest of the international organisations campaigning on this issue—Greenpeace—amounts to a mere fraction of 1% of the revenue of a corporation such as Monsanto,[2] while most of the campaigning has come from local groups or individuals with no real budget to speak of at all. As one of Monsanto's executives was reported to have said at a training conference, "These people work for nothing—how can you stop that?"[3]

Resistance has taken many forms, from the half million Indian farmers who marched carrying neem branches to protest against the World Trade Organisation and the patenting of life, to the woman from the UK who wrote to

Compelling evidence

On 29th March 1999, two women walked out of a court in Plymouth, UK, to be greeted by hundreds of cheering people. The crowd was told that the charges against the women, who freely admitted destroying a test site of genetically engineered maize, and who faced up to ten years in prison if found guilty, had been dropped. In a highly unusual admission, the prosecution said that "unfolding events", such as the swing in public opinion against genetic engineering, and political uncertainty about its future, influenced their decision not to pursue the case.[4]

Michael Schwarz, solicitor for the two women, described what happened: "We served the prosecution with ten expert reports detailing the risks posed by genetic engineering and the failures of the regulatory system. Within 24 hours I was told that the prosecution were going to drop all charges. It was made clear to me in a telephone call from the Crown Prosecution Service that this was a decision made at the highest level: by the Director of Public Prosecutions, David Calvert-Smith QC. This was a political—and in my experience, unprecedented—decision. By withdrawing the case from the jury the Crown have accepted that there was compelling evidence that the defendants had a lawful excuse to remove the GE maize. The last thing the Crown wanted was to see a jury—a microcosm of society—acquit people who admitted taking direct action against GE crops."[5]

Monsanto billing them for £6,418 and 82 pence for the trouble she has had to go to in order to avoid GE soya.[6]

"Diversity has characterised the campaign from its start," says Anne Ward, a local campaigner from Totnes, in Devon, England, where feelings ran high after a test site for genetically engineered maize was planted 275m from an organic farm. "Totnes is quite a fragmented community, made up of many disparate groups of people. So I have been amazed at the way genetic engineering has galvanised the community into action. I have been involved in campaigns in this town for years, but I have never seen so many people of all ages, political persuasions and class come together like this. I remember in particular one lady in her 80s who walked for three miles to get to a demonstration—it was the first one she had been to in her life."[7]

Similar experiences have been widely reported, not only on a local level, but also among established organisations that are beginning to form new and sometimes unusual alliances. As Zac Goldsmith commented in a recent editorial for *The Ecologist*, this "can only lead to admiration for Monsanto, who have single-handedly managed to unite a divided social and ecological movement."[8]

By 1999 major supermarkets, food producers, beverage companies, animal feed suppliers and restaurants across Europe and also in Japan, Australia, New Zealand, Korea, Thailand, the US, Canada, South Africa, Mexico, Brazil and Hong Kong, were bowing to pressure from the public, and beginning to exclude genetically engineered ingredients from their own-brand products. This even included corporations such as Unilever, the largest processed food producer in the world, which belongs to the multinational consortium EuropaBio. Just two years after EuropaBio hired PR company Burson Marsteller to help convince the public of the benefits of genetic engineering, Unilever was forced to withdraw genetically engineered soya from its foods in the UK, after consumer boycotts reduced sales of its product 'Beanfeast' by more than 50%.[9]

Geoff Lancaster, communications chief for British Sugar, a company which announced in 1998 that it would not use any GE sugar beet in its processing plants, gave his overview of the situation: "Our original line was that as long as the industry abided by all the statutory controls, we were OK. That no longer applies. Media and public attitudes have hardened following the soya debate . . . I realise this paints a rather bleak future for [GE] varieties. We are now at a crossroads."[10]

Another of the ways in which individuals and communities have been opposing the introduction of genetically engineered food and crops has been the establishment of 'GE-free zones'. Sometimes this has been a government decision, as in 1997 when Austria and Luxembourg defied the European Commission by banning Novartis' insect and herbicide resistant maize,[11] and when the Governor of Rio Grande del Sul, Brazil's major soybean growing state, declared the state a GE-free zone in March 1999.[12] More commonly, communities have been working from the ground up, trying to influence affairs at a local level. An example of this is the decision of the Waiheke Community Board in New Zealand to encourage all businesses and individuals residing or operating within the region to avoid the use of genetic engineering in every aspect of farming and food production.[13]

Similar initiatives have been undertaken in the UK, where people have been persuading shops and restaurants to source GE-free supplies, and where parents and teachers have urged local councils to remove genetically engineered ingredients from school meals. This has been very successful, and in February 1999 the sixty voting members of the Local

Government Association made a unanimous decision to recommend councils across the country to remove GE food from all their outlets—schools, town halls, and residential homes for the elderly. This recommendation affects about 500 councils and almost 10 million children in 26,000 schools in England and Wales, as well as 1.5 million local government workers. Although it is advisory rather than legally binding, by the time the LGA made their recommendation many councils in England and Wales had already started to remove genetically engineered food from school menus, and with sustained pressure from the public, many more plan to follow suit.[14]

Support for these initiatives has also come from the UK's leading chefs and food writers, more than a hundred of whom pledged to oppose the use of GE food, and to encourage other chefs and restaurants to do the same. A similar campaign is being run by the Euro-Toques, a 2,500 strong association of the top chefs in Europe. In a statement made in December 1998, the UK chefs and food writers said: "As food professionals we object to the introduction of [GE] foods into the food chain. This is imposing a genetic experiment on the public, which could have unpredictable and irreversible adverse consequences."[15] It was not just the top restaurateurs who objected, however. In December 1999 there were red faces at the headquarters of Monsanto UK when Sutcliffe Catering, the company responsible for providing food on the premises, announced that it, like thousands of other food suppliers in the country, was taking the decision to remove GE soya and maize from all the food served there.[16]

By 1999 there were hundreds of organisations around the world demanding, at the very least, a moratorium on genetic engineering in food and farming. These ranged from conservation groups, to organisations such as the British Medical

Association, third world groups to organisations such as COAG, which represents about 200,000 Spanish farmers.[17]

In the UK, by 1999 concerns had grown to such an extent that the government was having difficulties finding enough farmers who would be prepared to take part in the large-scale trials of GE crops that had been proposed. Amongst those who declined the invitation were two of Britain's biggest farmers, the Co-operative Wholesale Society and the Church Of England.[18]

Even when land was found for the trials, the crops were often pulled up, sometimes by the landowners themselves. Captain Fred Barker, for example, who enthusiastically volunteered to be the first farmer in the UK to take part in 'government' trials being conducted by AgrEvo, was subsequently instructed by the trustees of his family estate to destroy the GE oilseed rape being grown.[19] On a Sunday afternoon a few weeks later, another government trial was destroyed after more than 500 people dressed in white paper suits pulled up 25 acres of GE oilseed rape after a rally in Watlington, Oxfordshire.[20]

In the southern Indian state of Karnataka, the main farmers' association launched a campaign in the winter of 1998 called 'Operation Cremate Monsanto', uprooting and burning field trials of genetically engineered cotton, which they believed to have been planted illegally.[21] By December 1999, GE crops had also been destroyed in the US, Germany, France, Netherlands, Ireland, India, New Zealand, Australia, Brazil, and Greece.[22]

A number of legal cases have also been influential. Following a case brought by a coalition of environmental organisations in France in 1998, the Conseil d'Etat (the highest administrative court in the country) suspended authorisation to grow Novartis' GE maize in France. The French government went on to introduce a two-year moratorium on the commercial cultivation of genetically engineered crops that have wild relatives in Europe, such as oilseed rape and sugar beet.[23] And in June 1999, France, Italy, Denmark, Greece and Luxembourg announced that they would block any attempt to approve new varieties of genetically engineered crops in the EU for at least 18 months.[24]

In another case, a number of farming organisations, including the US National Family Farm Coalition, have joined together with environment and development groups around the world to launch a multi-billion dollar lawsuit against the biotech industry, alleging that companies such as Monsanto had "forced genetically-modified seeds onto the market at fixed prices without sufficient testing for safety to human health and the environment".[25]

"At each point in this project we keep thinking that we have reached the low point and that public thinking will stabilise," reflected polling adviser Stanley Greenberg in his report for Monsanto, "but we apparently have not reached that point. The latest survey shows a steady decline over the year, which may have accelerated in the most recent period."[26]

As outlined in this chapter, the biotech industry is beginning to face serious difficulties. However, regulatory hurdles delaying the commercialisation of GE products, the lack of market acceptance, and the near ruin of a particular company do not in themselves present the industry with insurmountable obstacles. Most governments emphasise that any delays faced by the industry are temporary; if an individual biotech company is ruined, its place will soon be taken by another; and with enough money and power, markets can be made. What is perhaps more of a lasting threat to the industry, is that increasing numbers of people are questioning the way we produce our food, the global economic system, and the power given to multinational corporations, and are willing to actively engage with the issues in whatever way they can.

'Genetic engineering' is not just a laboratory technique. It is a tool shaped by a particular worldview, supported by a particular political and economic framework. Some suggest that to challenge genetic engineering is to stand in the way of scientific progress—but the nature of progress depends on your point of view. I would argue that further threats to biodiversity and human health, potentially irreversible forms of pollution, expropriation of resources from Third World countries, increased corporate control of our food chain, the continued industrialisation of our agricultural systems, and the patenting of living organisms do not constitute progress.

Real progress will not come about through increased control and manipulation of the world for the short-term gains of the few. It will come about from the growing participation in social justice movements, the practice of a variety of sustainable agricultures, actions taken to protect the environment, and all ways of life that arise from an understanding that whatever we do to the world we do to ourselves. If we really want to bring about a society that respects life in all its wondrous diversity, then we can—but only if everyone who cares is also willing to act.

> "Small actions and choices can have major, although unpredictable, effects in determining what comes next. Among the possibilities is that the thousands of experiments and millions of choices to live more consciously will coalesce into a new civilisation that fosters community, provides possibilities for meaning, and sustains life for the planet."
>
> —Sarah van Gelder [27]

Resources

There are hundreds of groups around the world campaigning on genetic engineering issues. While some choose to focus on the genetic engineering of crops, others focus on patenting. Some want complete bans, some the labelling of genetically engineered products, and others moratoriums, while others are simply concentrating on raising public awareness in any way they can. Listed below are just a few of these groups and organisations—if you don't find the information you are seeking here, many of these groups may be able to put you in touch with other sources of information and support.

Points of contact in North America

The Edmonds Institute, Beth Burrows, 20319-92nd Avenue West, Edmonds, WA 98020. Tel: (425) 775-5383, Fax: (425) 670-8410.
<beb@igc.org> <www.edmonds-institute.org>
Conducts research, publishes policy analysis and scientific thought pieces, distributes information, sponsors public workshops, and provides expert witnesses at national events and for international bodies engaged in decision making. Disseminates information about and criticism of technology assessment, encourages *pro bono* research and policy analysis by scientists and scholars, and seeks to create alliances and coalitions with like-minded organisations and individuals. Can put people in touch with other campaigns.

RAFI International Office: 110 Osborne St., Suite 202, Winnipeg, MB R3L 1Y5, Canada. Tel: (204) 453-5259, Fax: (204) 925-8034.
USA office: PO Box 640, Pittsboro, NC 27312.
Tel: (919) 542-1396, Fax: (919) 542-0069. <rafi@rafi.org> <www.rafi.org>
An international non-governmental organisation dedicated to the conservation, sustainability and improvement of agricultural biodiversity, and to the socially responsible development of technologies useful to rural societies. RAFI is an important contact for info on patenting, terminator technology, the biotech industry and the loss of genetic diversity, and the relationship of these issues to human rights, agriculture and world food security.

Union of Concerned Scientists, National HQ, 2 Brattle Square, Cambridge, MA 02238-9105. Tel: (617) 547-5552.
<ucs@ucsusa.org> <www.ucsusa.org/agriculture/biotech.html>
Alliance of 70,000 committed citizens and leading scientists who aim to "augment rigorous scientific research with public education and citizen advocacy to help build a cleaner, healthier environment and a safer world". Provides a critique of the various applications of genetic engineering, and supports sustainable alternatives. See the Gene Exchange, p.129.

Alliance for Bio-Integrity, Midwest Office: PO Box 110, Iowa City, IA 52244-0110. Tel: (515) 472-5554, Fax: (515) 472-6431. <info@bio-integrity.org> <www.bio-integrity.org>
Organized a lawsuit against the FDA for its policy on GE food. Working with scientists, public interest organizations, and people from diverse faiths.

Bioengineering Action Network, PO Box 11703, Eugene, Oregon 97440. Tel: (541) 302-5020. <ban@tao.ca> <www.tao.ca/~ban>
Network of anti-GE activists in North America. Visit website to get contact details of local groups. For regular updates about direct action and other anti-GE info, email <majordomo@tao.ca> with 'subscribe ban' in the text.

BioDemocracy (Formerly Campaign for Food Safety), 860 Highway 61, Little Marais, MN 55614. Tel: (218) 226-4164, Fax: (218) 226-4157. <alliance@mr.net> <www.purefood.org/index.htm>
Dedicated to building a healthy, safe, and sustainable system of food production and consumption. Acts as a clearinghouse for info and grassroots technical assistance. Active and well informed on GE issues.

The Campaign to Label Genetically Engineered Foods, PO Box 55699, Seattle, WA 98155 Tel: (425) 771-4049, Fax: (603) 825-5841
<label@thecampaign.org> <www.thecampaign.org>
Consumer campaign lobbying Congress and the President to pass legislation that will require the labelling of GE foods in the US.

Center for Ethics and Toxics, 39141 S Highway 1, PO Box 673, Gualala, CA 95445. Tel: (707) 884-1700. <www.cetos.org>
Supports agricultural practices that are free from reliance on pesticides and encourages communities to declare and sustain 'Toxic Free Zones'. Director and Research Associate are co-authors of *Against the Grain* (see p. 132).

Center for Food Safety, 310 D Street NE, Washington, DC 20002. Tel: (202) 547-9359 Fax: (202) 547-9429.
<office@icta.org> <www.centerforfoodsafety.org>
Established to carry out "vital initiatives in food safety, environmental protection and sustainable agriculture". Aims to ensure the testing and labelling of GE foods. Has filed lawsuits against the US government on GE issues.

Consumers Union, 101 Truman Ave, Yonkers, NY 10703. Tel: (914) 378-2452. <hansmi@consumer.org> <www.consumersunion.org/>
Contact: Michael Hansen & Jean Halloran.
Active nationally and internationally on rBST and other GE food safety issues.

Corporate Watch, PO Box 29344, San Francisco, CA 94129. Tel: (415) 561-6568. <corpwatch@igc.org> <www.corpwatch.org>
Aims to document the impacts of transnational companies and supports initiatives for human rights, environmental justice and corporate accountability. Working to harness the internet as a vehicle for activism.

Council for Responsible Genetics, 5 Upland Road, Suite 3, Cambridge, MA 02140. Tel: (617) 868-0870, Fax: (617) 491-5344.
<marty@gene-watch.org> <www.gene-watch.org>
Focuses on human genetics issues. Also active on biosafety and consumer 'right to know' issues. Produces and distributes educational materials.

Earth First!, POB 1415, Eugene, OR 97440. Tel: (541) 344-8004.
<earthfirst@igc.org> <www.enviroweb.org/ef/contact.html#USA>
Network of autonomous local groups. Activities include grassroots organising,
civil disobedience and direct action.

The Foundation on Economic Trends, 1660 L Street NW, Suite 216,
Washington DC 20036. Tel: (202) 466-2823, Fax: (202) 429-9602.
<jrifkin@foet.org> <www.biotechcentury.org>
Examines emerging trends in science and technology and their impacts on the
environment, the economy, culture and society. The President of the
Foundation is Jeremy Rifkin, author of *The Biotech Century* (see p.132).

Greenpeace USA, Charles Margulis, c/o Greenpeace GE Campaign, 1817
Gough Street, Baltimore, MD 21231. Tel: (410) 327-3770.
<www.greenpeaceusa.org> <charles.margulis@dialb.greenpeace.org> GE
contact for Greenpeace in the US. See Greenpeace International p.128.

Institute for Agriculture and Trade Policy, 2105 1st Avenue South,
Minneapolis, MN 55404. Tel: (612) 870-0453. <www.iatp.org>
Aims to create environmentally and economically sustainable communities and
regions through sound agriculture and trade policy. Provides educational materials and technical assistance, and works to build networks.

Institute for Social Ecology, 1118 Maple Hill Road, Plainfield, VT 05667.
Tel: (802) 454 8493. <ise@sover.net> <http://ise.rootmedia.org>
Educational and activist organization. Offers summer intensives, workshops and
colloquia on a variety of ecological issues, including the science and politics of
GE. Seeks a future of grassroots democracy, and moral economies that seek to
reharmonize human communities with the natural world.

International Forum on Globalization, 1555 Pacific Avenue, San
Francisco, CA 94109. Tel: (415) 771-3394. <ifg@ifg.org> <www.ifg.org>
An alliance of sixty activists, scholars, economists, researchers, and writers
formed to stimulate new thinking, joint activity, and public education in
response to the rapidly emerging economic and political arrangement called
the global economy. Represents 40 organisations in 20 countries.

Mothers & Others for a Livable Planet, 40 West 20th Street, New York,
NY 10011-4211. Tel: (212) 242-0010.
<Mothers@mothers.org> <www.mothers.org/mothers/>
Educates consumers about benefits of sustainable farming to health and environment, and encourages them to use their 'buying power' to support farmers who
are moving towards safer, sustainable production methods.

Mothers for Natural Law, PO Box 1177, Fairfield, IA 52556.
Tel: (515) 472-2809. <www.safe-food.org>
Focuses on GE issues. Current activities include: a public awareness campaign on
the dangers of GE foods, sourcing supplies of GE-free food, promoting 'GE-free'
certification, and keeping GE out of the organic market.

National Family Farm Coalition, 110 Maryland Ave NE, Washington, DC
20002. Tel: (202) 543-5675.
Leading anti-GE farm group in the US. Hundreds of small, family farm member groups throughout the US (mostly in the mid-west and NE).

Native Forest Network, NFN/ACERCA, PO Box 57, Burlington, VT 05402.
Tel: (802) 863-0571. <nfnena@sover.net> <www.nativeforest.org>
Working to protect native forests worldwide and looking at the connections between GE and forestry (e.g. the transnational efforts to bring GE eucalyptus plantations into the rainforest of Chiapas in southern Mexico).

Northeast Resistance Against Genetic Engineering (NERAGE),
Vermont Clearinghouse, c/o Institute for Social Ecology (see p.125).
Tel: (802) 454-9957. <ise@sover.net> <www.bckweb.com/nerage>
Activist network of individuals and groups concerned about GE issues. Committed to exposing the implications of GE for our health and the environment, as well as for the future of agriculture and our social relationships, through public education campaigns and direct action.

Organic Consumers Association, 860 Highway 61, Little Marais,
MN 55614. Tel: (218) 226-4792, Fax: (218) 226-4157.
<oca@purefood.org> <www.organicconsumers.org>
Protects the integrity of the organic label, promotes sustainable agriculture, and opposes the use of genetic engineering in food and farming.

Pesticide Action Network (PAN), North American Office (PANNA),
49 Powell St., Suite 500, San Francisco, CA 94102. Tel: (415) 981-1771,
Fax: 415 981 1991. <panna@panna.org> <www.panna.org/panna>
Has campaigned to replace pesticides with ecologically sound alternatives since 1982. PANNA is one of five PAN Regional Centers in Africa, Asia/Pacific, Latin America, Europe and North America.

The American Community Gardening Association (ACGA),
100 N. 20th Street, 5th Floor Philadelphia, PA 19103-1495.
Tel: (215) 988-8785, Fax: (215) 988-8810.
<smccabe@pennhort.org> <www.communitygarden.org/>.
Works to promote and support all aspects of community gardening, urban forestry, and preservation and management of open space. Facilitates the formation and expansion of state and regional community gardening networks, encourages research and conducts educational programs.

Other international and national organisations

ActionAid, Campaigns, Hamlyn House, Macdonald Road, Archway, London
N19 5PG, UK. Tel: +44 207 561 7611, Fax: +44 207 281 5146.
<campaigns@actionaid.org.uk> <www.actionaid.org>
Active at a grass roots level in 30 countries. Working on an international food rights campaign, and is increasingly focused on GE and patenting. Published a report on Astra-Zeneca in May 99—see their website.

ANPED, PO Box 59030, 1040 KA Amsterdam, The Netherlands. Tel: +31 20
4751742, <anped@anped.antenna.nl> <www.antenna.nl/anped>.
Tel/Fax: 0208 672 3454. <iza@cpa-iza.u-net.com> GE contact: Iza Kruszewska.
Network of local and national groups from Central-Eastern (CEE) and Western

Europe, Newly Independent States (NIS) and North America, working on issues of production and consumption, sustainable development and participatory democracy. Their activities on GE food and agriculture are focused on building a regional network of groups in CEE and NIS, starting campaigns in those countries where none exist, and providing information and other support to other groups already campaigning on this issue.

ASEED Europe, PO Box 92066, 1090 AB Amsterdam, The Netherlands. Tel: +31 20 66 82 236, Fax: +31 20 66 50 166.
<aseedeur@antenna.nl> <www.antenna.nl/aseed/>
Activist network committed to ecological and social justice, based in over 30 European countries. Campaigning on GE and food security since 1996. Develops information tools on GE, such as corporate profiles of the main players in the industry, and initiates and coordinates actions and campaigns.

Australian Gene Ethics Network, 340 Gore St, Fitzroy 3065, Victoria, Australia. Tel: +61 3 9416 2222, Fax: +61 3 9416 0767.
<acfgenet@peg.apc.org> <www.zero.com.au/agen/>
Federation of groups and individuals in Australia promoting critical discussion and debate on the environmental, social and ethical impacts of GE.

Compassion in World Farming, Charles House, 5A Charles Street, Petersfield GU32 3EH, UK. Tel: +44 1730 264208 / 268863, Fax: +44 1730 260791. <compassion@ciwf.co.uk> <www.ciwf.co.uk>
Concerned about the genetic engineering of animals and xenotransplantation. Provides videos and other resources suitable for use in schools, etc.

Consumers International, 24 Highbury Crescent, London, N5 1RX, UK. Tel: +44 207 226 6663, Fax: +44 207 354 0607. <consint@consint.org>
<www.oneworld.org/consumers//campaigns/food/>
Federation of 245 consumer organisations in 110 countries. Active on GE issues internationally, e.g. labelling negotiations at Codex Alimentarius.

The Gaia Foundation, 18 Well Walk, Hampstead, London NW3 1LD, UK. Tel: +44 207 435 5000, Fax: +44 207 431 0551. <gaiafund@gn.apc.org>
Links with farmers, scientists and grass roots organisations in Third World. Raises awareness about the impact of GE and patenting in these countries.

Genetic Engineering Network, PO Box 9656, London N4 4JY, UK. Tel: +44 208 374 9516.
<info@genetix.freeserve.co.uk> <www.visitweb.com/totnes>
Links grass-roots GE campaigns in the UK with other national and international campaigns. Aims to provide 'information for action', e.g. how to set up a local campaign. Publishes *Genetix Update*, a bi-monthly newsletter.

Genetic Resources Action International (GRAIN), Girona 25, pral., E-08010 Barcelona, Spain. Tel: +34 93 301 1381, Fax: +34 93 301 1627.
<grain@bcn.servicom.es> <www.grain.org>
International NGO with offices in Spain and the Philippines, established in 1990 to help further a global movement of popular action against the threat of genetic erosion. Very well informed on patenting, biodiversity, etc.

genetiX snowball, One World Centre, 6 Mount Street, Manchester, M2 5NS. Tel: +44 161 834 0295, Fax: +44 161 834 8187. <genetixsnowball@onet.co.uk> <www.gn.apc.org/pmhp/gs/>
A 'campaign of nonviolent civil responsibility' aiming to build active resistance to GE. Visit their website to download the genetiX snowball handbook.

GeneWatch UK, The Courtyard, Whitecross Road, Tideswell, Buxton SK17 8NY, UK. Tel: +44 1298 871898, Fax: +44 1298 872531. <gene.watch@dial.pipex.com>
Specialises in the science, ethics, risks and regulation of GE. Undertakes research and analysis on its implications. Published a detailed report in June 1999 on the environmental releases of GE microorganisms.

Greenpeace International, Chaussesstr. 131-10115 Berlin, Germany. Tel: +49 30 30 889914, Fax: 889930. <www.greenpeace.org/~geneng>
International environmental organisation that both lobbies and takes non-violent direct action. Their new GE website includes info on a range of issues, as well as press releases, details about actions, etc.

Research Foundation For Science, Technology & Natural Resource Policy, A-60 Haus Khas, New Delhi 110016, India. Tel: +91 11 696 8077, Fax: 685 6795. <twn@uvn.ernet.in> <www.indiaserver.com/betas/vshiva>
Focuses on biodiversity conservation, food security, globalisation, patenting, genetic engineering, biosafety, sustainable agriculture, WTO and GATT.

Third World Network, International Secretariat, 228 Macalister Road, 10400 Penang, Malaysia. Tel: +60 4 2266728 or 2266159, Fax: 2264505. <twn@igc.apc.org> <www.southside.org.sg/souths/twn/twn.htm>
Network of organisations and individuals involved in issues relating to development, the Third World and North-South issues. Their website is a useful source of information about biopiracy, patents, the WTO and GE.

Women's Environmental Network, 87 Worship Street, London EC2A 2BE, UK. Tel: +44 207 247 3327, Fax: +44 207 247 4740. <testtube@gn.apc.org> <www.gn.apc.org/wen>
WEN has a core group of geneticists and biologists, and is very active and well informed on GE issues. Information packs available with detailed information about GE-related issues, and also practical campaigning resources with ideas for action that can be taken by individuals and groups.

Magazines, Journals etc

Corner House Briefings, The Corner House, PO Box 3137, Station Road, Sturminster Newton, Dorset DT10 1YJ, UK. <cornerhouse@gn.apc.org> <www.icaap.org/Cornerhouse/articles.html>
Detailed briefings by a skilled team of researchers, include ones on patenting, GE and World Hunger, and cloning.

The Ecologist, Unit 18, Chelsea Wharf, 15 Lots Road, London SW10 0QJ, UK. Tel: +44 207 351 3578, Fax: 351 3617. <www.gn.apc.org/ecologist>
Publishing radical green thought for 30 years. Regular features and updates on issues related to genetic engineering and the biotech industry.

Earth First! Journal, Contact: Earth First!, POB 1415, Eugene, OR 97440, USA. Tel: (541) 344-8004, Fax: (541) 344-7688.
<earthfirst@igc.org> <www.enviroweb.org/ef>
A forum for environmental activists: news, views and discussions of direct action. Published eight times a year.

The Gene Exchange <www.ucsusa.org/publications/index.html>
A valuable resource, edited by Jane Rissler and Margaret Mellon from the Union of Concerned Scientists. Look up website for info on how to receive by email. By post, send request to Direct Mail Administrator at UCS (see p.123).

GenEthics News, PO Box 6313, London N16 0DY, UK.
<101561.3476@compuserve.com>
<http://ourworld.compuserve.com/homepages/genethicsnews/>
Founded by geneticist Dr. David King. Bimonthly newsletter covering the ethical, social and environmental issues raised by GE and human genetics.

GeneWATCH (not to be confused with GeneWatch UK!): a magazine produced by the Council for Responsible Genetics (see p.124).

GeneWatch Briefings Valuable briefings from GeneWatch UK (see p.128).

Manual for Assessing Ecological and Human Health Effects of Genetically Engineered Organisms A two-volume, peer-reviewed manual written by a group of scientists from a wide range of disciplines. Available for cost of mailing from the Edmonds Institute (see p.123).

Of Cabbages And Kings Anti-GE cartoon book available for cost of mailing from ASEED. Can also be downloaded from their website (see p.127).

The Permaculture Activist, PO Box 1209W, Black Mountain, NC 28711.
<pcactiv@metalab.unc.edu> <http://metalab.unc.edu/pc-activist>
Published 3 times a year. Contact for information about permaculture in US.

The American Permaculture Directory Contact Shade Tree Publishing, 5515 N 7th St., Suite 5144, Phoenix, AZ 85014. Tel: (602) 279-3713.
<www.permaculture.net/APD.html> Info on courses, activists and groups.

RAFI Communiqués Valuable and detailed analysis of patenting issues, terminator technology and the life science industry. See RAFI, p.123.

The Ram's Horn, S6 C27 RR#1, Sorrento, BC, V0E 2W0, Canada.
Tel/Fax: (250) 675-4866. <ramshorn@ramshorn.bc.ca>
Eight pages of information and analysis of the food system, published monthly. Increasingly focused on the issues and dangers posed by GE.

Seedling Quarterly newsletter published by GRAIN (see p.127). An exchange of news and analysis among those engaged in GE issues.

Selling Suicide: farming, false promises and genetic engineering in developing countries Report published by Christian Aid and available from them at PO Box 100, London SE1 7RT, UK, or on the web at <www.christian-aid.org.uk/reports/suicide/index.html>.

The Splice of Life c/o The Genetics Forum, 94 White Lion Street, London N1 9PS. Tel: +44 207 837 9229, Fax: +44 207 837 1141.
geneticforum@gn.apc.org <www.geneticsforum.org.uk>
The magazine produced by The Genetics Forum, a UK watchdog on GE issues. It covers the social, environmental and ethical implications of GE.

Email information services

BAN GEF Free US-run list with useful daily digest. For instructions send email to <Ban-GEF@lists.txinfinet.com> with "HELP" in the Subject line.

BIO-IPR In-depth list put out by GRAIN (see p.127). Circulates information about recent developments in the field of patenting related to biodiversity, etc. To join, send the word "subscribe" (no quote marks) as the subject of an email message to <bio-ipr-request@cuenet.com>.

Canadian Biotech listserv Window on Canadian biotech activities and a communications link for Canadian activists. To subscribe, send a brief message including information about yourself to <jest-west@sfu.ca>.

Center for Food Safety To subscribe to Campaign for Food Safety News (formerly called Food Bytes), send an email to <majordomo@mr.net> with "subscribe pure-food-action" in the body of the text.

Genetic Engineering Network (GEN) List 1 A UK-based, free service; a moderated list averaging about 5 emails per day. Includes information about all aspects of GE with a close eye on worldwide resistance to the technology (Archived as 'info4action' at <www.gene.ch/> since 1998). **GEN List 2** is a much quieter list which sends out GEN's newsletter as well as action reports. Email <genetics@gn.apc.org> to subscribe and state which of the two lists you want to be on (all info in list 2 is covered by list 1).

GENET Moderated list: information exchange among European NGOs and grassroot groups (archived at <www.gene.ch/> since 1998). Email <genet@agoranet.be> for subscription details.

GENTECH Unmoderated list about all aspects of genetic engineering (archived at <www.gene.ch> since 1995). Send a message with the word "subscribe" in the subject to: <GENTECH-REQUEST@tribe.ping.de>.

Nginews Very useful e-mail bulletins from Norfolk Genetic Information Network. Email <mail@icsenglish.com> with the word "subscribe".

NLP Wessex List Concentrates on crop issues. Subscribe on website: <www.btinternet.com/~nlpwessex/Documents/gmocarto.htm>.

Organic Consumers Association To subscribe to the free electronic newsletter, *Organic View*, send an email to <organicview@organic-consumers.org> with "subscribe" written in the body of the text.

PANUPS Free weekly on-line news service from PANNA (see p.126). Subscription information available on website <www.panna.org/panna>.

Rachel's Environment & Health Weekly Weekly news with searchable archive at <www.monitor.net/rachel/>. Send the word "Subscribe" by itself (no quote marks) in an email to: <rachel-weekly-request@world.std.com>.

Other information on websites

Ag BioTech InfoNet <www.biotech-info.net/about.html> Focuses on scientific reports and technical analysis of GE issues. Aims to provide a forum where a broad spectrum of people and organizations can raise tough questions, report new technical findings, and offer conflicting views.

Biodynamics <www.biodynamics.com> Website of the Biodynamic Association of America. Provides information about biodynamic agriculture, which seeks to work with the forces of nature without chemicals.

Food 'n' Health 'n' Hope <www.seizetheday.org/main.htm> Visit this site to listen to (or download) a song about a certain company.

Genetic Engineering and Its Dangers
<http://userwww.sfsu.edu/~rone/gedanger.htm> À series of articles and essays which covers a wide range of issues including GE and biological weapons, spiritual perspectives etc. Compiled by Dr Ron Epstein from the Philosophy Department of San Francisco State University.

Hexterminators <www.artactivist.com> Imaginative San Francisco-based activists who use art and street theater to highlight the hazards of GE and promote organic alternatives. Ideas for action, GE info, resources, etc.

Ifgene <www.anth.org/ifgene/articles.htm> Particular focus on the worldviews with which people approach science, and the moral and spiritual implications of biotechnology.

One World Online <www.oneworld.org/guides/biotech/front.html> Dedicated to promoting human rights and sustainable development by harnessing the democratic potential of the Internet. Highly regarded website.

Organic Farming Research Foundation <www.ofrf.org> Website of an organisation which sponsors research into organic farming practices.

Physicians and Scientists for Responsible Application of Science and Technology <www.psagef.org/indexgen.htm> Website has a range of science-based information about the hazards of GE.

Primal Seeds <www.primalseeds.org> Primal Seeds exists as a network for those who wish to actively engage in protecting biodiversity and creating local food security. It aims to help people create an alternative vision of agriculture through actions, discussions, skill sharing, and ideas. The website offers a wealth of news, ideas for actions, and other information e.g. on seeds and seed exchanges, agriculture, corporate control, biopiracy, permaculture, wild plants, herbs, and urban gardening etc.

Totnes Genetics Group Website A grassroots campaign in the UK with a colourful and informative website. Useful ideas for local campaigning.

USDA Animal and Plant Health Inspection Service (APHIS) <www.aphis.usda.gov/bbep/bp/> Find out details here of approvals granted by the USDA, EPA and FDA for genetically engineered organisms.

Recommended books

World Hunger: Twelve Myths by Frances Moore Lappé, Joseph Collins and Peter Rosset, Grove Press, New York, 1998. Excellent exposé of the myths that prevent us from effectively addressing the problem of world hunger.

The Biotech Century: Harnessing the Gene and Remaking the World by Jeremy Rifkin, Tarcher/Putnam (New York), 1998. Very readable overview of developments within the field of GE.

The Ecological Risks of Engineered Crops by Jane Rissler and Margaret Mellon, MIT Press, 1996. One of the key texts on GE and the environment.

Biotechnology, Weapons and Humanity, British Medical Association, London, 1999. Covers the issue of GE and biological warfare.

Biopiracy: The Plunder of Nature and Knowledge by Vandana Shiva, Green Books, 1998. Patenting, biopiracy and the 'new colonialism'.

Exploding the Gene Myth by Ruth Hubbard and Elijah Wald, Beacon Press, 1997. A critique of genetic determinism.

Brave New Worlds: Staying Human in the Genetic Future by Bryan Appleyard, Viking Press, New York, 1998. Explores human GE issues.

Farmageddon: Food and the Culture of Biotechnology by Brewster Kneen, New Society, Gabriola Island, BC, 1999. Critique of GE as reductionist science, motivated by corporate profit.

Redesigning Life compiled by Brian Tokar, Zed Books, London, 2000. Essays from over twenty critics of GE. Recommended.

Against the Grain by Mark Lappé and Britt Bailey, Earthscan, 1999. Covers agricultural GE issues, such as the impacts of herbicide-resistant crops.

Eat Your Genes: How Genetically Modified Food is Entering our Diet by Stephen Nottingham, Zed Books Ltd, 1998. Detailed information on issues ranging from the science of GE to the regulatory systems in Europe & USA.

The Human Body Shop: The Engineering and Marketing of Life by Andrew Kimbrell, HarperSanFrancisco, 1993. Accessible introduction to the commercialisation of the human body.

For all those who are interested in the relationship between apathy and action, and the response of the human heart to ecological crisis, I recommend the following two books:

Coming Back to Life: Practices to Reconnect Our Lives, Our World by Joanna Macy and Molly Young Brown, New Society, Canada, 1998. Contains a wealth of practical information and exercises that can be used by both groups and individuals.

World as Lover, World as Self by Joanna Macy, Parallax Press, Berkeley, 1991. Explores activism, ecological despair, systems theory and spiritual practice.

References

The quotation by Wendell Berry on the page facing the Introduction is from his essay 'Conservation is Good Work', published in his book *Sex, Economy, Freedom and Community*, Pantheon Books, New York and San Francisco, 1992.

Introduction

1. John Losey, et al., 'Transgenic pollen harms monarch larvae', *Nature* 399: 214, 20 May 1999.
2. Frank Mitsch and Jennifer Mitchell, 'Ag Biotech: Thanks, But No Thanks?', Deutsche Bank, Alex. Brown, 12 July 1999. <www.biotech-info.net/Deutsche.html>; 'UK Supermarkets Move Out of GM-fed Animal Products', Greenpeace UK press release, 20 December 1999; Ronnie Cummins & Ben Lilliston, 'Global Resistance Mounts Against Monsanto & Genetic Engineering', *Food Bytes* No. 17, 2 March 1999 <www.purefood.org>; Ronnie Cummins & Ben Lilliston, *Campaign for Food Safety News* No. 22, 21 October 1999 <www.purefood.org>.
3. Personal communication on US agricultural exports from the USDA Foreign Agricultural Service to Stokely Webster, Greenpeace International, 23 November 1999.
4. 'GM crop warning for US farmers', BBC News Online, 24 November 1999 <http://news2.thls.bbc.co.uk/hi/english/world/americas/newsid_535000/535387.stm>.

Chapter 1 What is Genetic Engineering?

1 Quoted from Regine Kollek, 'The gene—that obscure object of desire', from Miges Baumann et al, *The Life Industry*, IT Publications, London 1996.
2 Fritjof Capra, *The Web of Life*, HarperCollins, 1996, Chs. 9 & 10.
3 Michael Antoniou, 'Genetic Engineering and Traditional Breeding Methods: A Technical Perspective', Physicians and Scientists for Responsible Applications of Science and Technology <www.psrast.org/mianbree.htm>.
4 Quoted in Evelyn Fox Keller, *Love, power and learning* (Liebe, Macht und Erkenntis), Hanser: Munich, 1986, p.179.
5 Mae-Wan Ho, 'The Unholy Alliance', *The Ecologist*, Vol.27, No.4, July/August 1997, p.153 <www.purefood.org/ge/unholya.html>.
6 R. Hightower et al, 'Expression of antifreeze proteins in transgenic plants', *Plant Molecular Biology* Vol.17, 1991, pp.1013-21; Michael Antoniou, 'Breaking the Chain', *Living Earth*, Vol.197, Jan-Mar 1998, p.20.
7 Jeremy Rifkin, *Declaration of a Heretic*, Routledge and Kegan Paul, London, 1985, p.53.
8 R. Steinbrecher and M.-W. Ho, 'Fatal Flaws in Food Safety Assessment: Critique of The Joint FAO/WHO Biotechnology and Food Safety Report', 1996. <www.psagef.org/fao96.htm>.
9 John Fagan, 'Assessing the safety and nutritional quality of genetically engineered

foods' <www.psagef.org/jfassess.htm>.

10 Beatrix Tappeser, 'Gutachten zur wissenschaften Zielsetzung und dem wissenschaftlichen Sinn des Freisetzungsexperimentes mit transgenen Petunien.' Oeko-Institut e.V., Freiburg, 1990; Ricarda Steinbrecher, 'Cotton Picking Blues', *The New Internationalist*, No.293, August 1997, p.22.

11 S. Hormick, 'Effects of a Genetically-Engineered Endophyte on the Yield and Nutrient Content of Corn', 1997. Interpretive summary available at: <www.geocities.com/Athens/1527/btcorn.html>.

12 T. Inose and K. Murata, 'Enhanced accumulation of toxic compound in yeast cells having high glycolytic activity: a case study on the safety of genetically engineered yeast', *Int. J. Food Science Tech.*, 30, 1995, pp.141-146.

13 Monsanto scientific publication in *Journal of Nutrition*, Vol.126, pp.717-727.

14 ACNFP Review (Application to the United Kingdom Advisory Committee on Novel Foods and Processes), page 59, table 7, re '3.5% corrected milk'.

15 'Making Crops Make More Starch', *BBSRC Business*, UK Biotechnology and Biological Sciences Research Council, January 1998, pp.6-8.

16 George Wald, '*The Case Against Genetic Engineering*', The Recombinant DNA Debate, Jackson and Stich, eds., pp.127-128. (Reprinted from *The Sciences*, Sept/Oct 1976).

17 'European Response to Genetically Modified Soybeans', Press Release from American Soybean Association, November 1996 <www.oilseeds.org/asa/news.htm>.

18 J. Fagan, 'Importation of Ciba-Geigy's Bt maize is scientifically indefensible', <www.netlink.de/gen/BTCorn.htm>.

19 A.N. Mayeno and G.J. Gleich, 'Eosinophilia-myalgia syndrome and tryptophan production: a cautionary tale', *TIBTECH*, Vol.12, 1994, pp.346-352.

20 L.A. Love et al, 'Pathological and immunological effects of ingesting l-tryptophan and 1,1'-ethylidenebis (l-tryptophan) in Lewis rats', *Journal of Clinical Investigation*, Vol.91, March 1993, pp.804-811; D.E. Brenneman et al, 'A decomposition product of a contaminant implicated in l-tryptophan eosinophilia myalgia syndrome affects spinal cord neuronal cell death and survival through stereospecific, maturation and partly interleukin-1-dependent mechanisms', *Journal of Pharmacology and Experimental Therapeutics*, Vol.266(2), 1993, pp.1029-1035.

21 Personal communication from William Crist, independent researcher from North Carolina, USA.

22 As 19; P. Raphals, 'Does medical mystery threaten biotech?' *Science*, Vol.249, 1990, p.619; P. Raphals, 'EMS deaths: Is recombinant DNA technology involved?', *The Medical Post*, Toronto, 6 November 1990; also 21.

23 P. Raphals, 'Does medical mystery threaten biotech?' *Science*, Vol.249, 1990, p.619.

24 As 23.

25 'Report on Riboflavin Derived from Genetically Modified (GM) *Bacillus subtilis* using Fermentation Technology', ACNFP Report, Ministry of Agriculture, Fisheries and Food Publications, 1996.

26 As 22.

27 A. Sloan and M. Powers, 'A perspective on popular perspections of adverse reactions to foods'. Journal of Allergy and Clinical Immunology, Vol.78, 1986, pp.127-133.

28 J. Nordlee et al, 'Identification of a brazil-nut allergen in transgenic soybeans', *The New England Journal of Medicine*, Vol.334(11), 1996, pp.688-692; V. Melo et al, 'Allergenicity and tolerance to proteins from Brazil nut (*Bertholletia excelsa* H.B.K.)', *Food Agric. Immunol.*, Vol.6, 1994, pp.185-195.

29 M. Nestle, 'Allergies to transgenic foods: Questions of policy', *The New England Journal of Medicine*, Vol.334(11), 1996, pp.726-727.

30 'Resistance to Antibiotics and other Antimicrobial Agents', 7th Report of the House of Lords Select Committee on Science and Technology, 1998.

31 Martyn Halle, 'A Pill to beat the Hospital Superbug', *Daily Mail*, 30 March 1999.

32 H. Tschäpe, 'The spread of plasmids as a function of bacterial adaptablity', *FEMS Microbiology Ecology*, Vol.15, 1994, pp.23-32; M.-W. Ho et al, 'Gene Technology and Gene Ecology of Infectious Diseases', *Microbial Ecology in Health and Disease*, Vol.10, 1998, pp.33-59.

33 *AgBiotech News and Information*, Vol.8 (9), pp.159N.

34 T. Hoffmann, C. Golz, and O. Schieder, 'Foreign DNA sequences are received by a wild-type strain of *Aspergillus niger* after co-culture with transgenic higher plants', *Current Genetics*, Vol.27, 1994, pp.70-76.

Chapter 2 GE and the Environment

1 C. James, 'Global Review of Commercialised Transgenic Crops: 1998', ISAAA Briefs No.8. ISAAA: Ithaca, NY, 1998.

2 M. Lappe and B. Bailey, *Against the Grain*, Common Courage Press (1998), pp.75-76.

3 Mark Arax and Jeanne Brokaw, 'No Way Around Roundup—Monsanto's bioengineered seeds are designed to require more of the company's herbicide', *Mother Jones*, Jan/Feb 1997 <www.purefood.org/mothjones.html>.

4 'Monsanto releases seed piracy case settlement details', Monsanto Press Release 12, September 1998.

5 Monsanto Press release, São Paulo, Brazil, 14 September 1998 <www.monsanto.com/Monsanto/mediacenter/98/98sep14RoundupBrazil.html>

6 'Sales Boom in Pesticides: Herbicide-Resistant Crops Expected to Benefit 1997 US Herbicide Sales', *The Gene Exchange—A Public Voice on Biotechnology and Agriculture*, Union of Concerned Scientists, Fall 1997 <www.ucsusa.org/Gene/F97.agribusiness.html>; 'World Agchem Market Recovery Continues', *Agrow*, No.284, 11 July 1997, pp.21-22.

7 'Genetically Engineered Oilseed Rape: Agricultural Saviour or New Form of Pollution?', GeneWatch Briefing No.2, May 1998, (The Courtyard, Whitecross Rd, Tideswell, Buxton, Derbyshire SK17 8NY, UK).

8 C. Cox 'Herbicide Factsheet: Glufosinate', *Journal of Pesticide Reform*, Vol.16 No.4, 1996, pp.15-19; also as 7.

9 Stephen Nottingham, *Eat Your Genes*, Zed Books, London, 1998, p.44.

10 'Glyphosate: Environmental Health Criteria 159', World Health Organisation, United Nations Environment Programme, International Labor Organisation, Geneva, Switzerland, 1994.

11 J.A. Springett and R.A.J. Gray, 'Effect of repeated low doses of biocides on the earthworm *Aporrectodea caliginosa* in laboratory culture', *Soil. Biol. Biochem.*, Vol.24(12), pp.1739-1744.

12 D. Estok, B. Freedman and D. Boyle (1989), 'Effects of the herbicides 2,4-D, glyphosate, hexazinone, and triclopyr on the growth of three species of ectomycorrhizal fungi', *Bull. Environ. Contam. Toxicol.*, Vol.42, pp.835-839; P. Chakravarty and L. Chatarpaul (1990), 'Non-target effect of herbicides: I. Effect of glyphosate and hexazinone on soil microbial activity, Microbial population, and in-vitro growth of ectomycorrhizal fungi', *Pestic. Sci.*, Vol.28, pp.233-241; P. Chakravarty and S.S. Sidhu (1987),

'Effects of glyphosate, hexazinone and triclorpyr on in vitro growth of five species of ectomycorrhizal fungi', *Eur. J. For. Path.*, Vol.17, pp.204-210; S.S. Sidhu and P. Chakravarty (1990), 'Effect of selected forestry herbicides on ectomycorrhizal development and seedling growth of lodgepole pine and white spruce under controlled and field environment', *Eur J. For. Path.*, Vol.20, pp.77-94.

13 C. Cox, 'Glyphosate, Part 2: Human Exposure and Ecological Effects, Herbicide Factsheet', *Journal of Pesticide Reform*, Winter 1995 Vol.15, No.4 (from the Northwest Coalition for Alternatives to Pesticides); W.S. Pease et al. (1993), 'Preventing pesticide-related illness in California agriculture: Strategies and priorities', Environmental Health Policy Program Report. Berkeley, CA: University of California. School of Public Health. California Policy Seminar.

14 C. Cox, 'Glyphosate, Part 1: Toxicology, Herbicide Factsheet', *Journal of Pesticide Reform*, Vol.15 No.3, Fall 1995 (from the Northwest Coalition for Alternatives to Pesticides).

15 Margaret Mellon, 'The Last Silver Bullet?', *The GeneExchange—A Public Voice on Biotechnology and Agriculture*, Union of Concerned Scientists, Winter 1996 <www.ucsusa.org/cgi-bin/AT-ucssearch.cgi>.

16 David Holzman, 'Agricultural Biotechnology: Report Leads to Debate on Benefits of Transgenic Corn and Soybean Crops', *Genetic Engineering News*, Vol.19 No.8, 15 April 1999, p.29.

17 'Monsanto Reduces Price Of US Roundup Herbicide Brands By $6 To $10/Gallon', Monsanto Press release, September 1998 <www.monsanto.com/monsanto/mediacenter/98/98sep1_RRprice.html>.

18 As 15.

19 J. Pratley et al., 'Glyphosate Resistance in Annual Ryegrass', *Proceedings of the 11th Conference*, Grasslands Society of New South Wales, 1996; D.S. Gill, 'Development of Herbicide Resistance in Annual Ryegrass Populations in the Cropping Belt of Western Australia', *Australian Journal of Exp. Agriculture*, Vol.3, 1995, pp.67-72; Jeannette Batz, 'The Eighth Day', *Riverfront Times*, St. Louis, MO, Dec. 11th, 1996 <www.purefood.org/eight.html>.

20 'Biotechnology and Pest Control: Quick Fix vs. Sustainable Control', *Global Pesticide Campaigner*, Vol.1, No.2, January 1991, pp.1 & pp.6-8.

21 As 15.

22 C. James, 'Global Review of Commercialised Transgenic Crops: 1998'. ISAAA Briefs No.8. ISAAA: Ithaca, NY, 1998.

23 B.E. Tabashnik, 'Evolution of Resistance to *Bacillus thuringiensis*', *Annual Review of Entomology*, Vol.39, 1994, pp.47-9. B.E. Tabashnik, Y-B. Liu, N. Finson, L. Masson, and D.G. Heckel, 'One gene in diamondback moth confers resistance to four Bt toxins', *Proceedings of the National Academy of Sciences*, USA, Vol.94, 1997, pp.1640-4.

24 EPA Pesticide Fact-Sheet 4/98, *Bacillus thuringiensis Cry IA(b) delta-endotoxin and the genetic material necessary for its production (Plasmid vector pclB 4431) in corn.* OPPTS, 1994.

25 Janelle Carter, 'Intoxicating Bacterium Kills Plants—EPA Is Sued Over Gene-Altered Crops', *Associated Press*, 18 February 1999.

26 A. Hilbeck, W.J. Moar, M. Pusztai-Carey, A. Filippini & F. Zigler, 'Toxicity of *Bacillus thuringiensis* CryIAb toxin to the predator *Chrysoperla carnea* (Neuroptera: Chrysopidae)'. *Environmental Entomology*, Vol.27, No.4, August 1998.

27 C. Tudge, '*The Engineer in the Garden. Genetics: From the Idea of Heredity to the Creation of Life*', Jonathan Cape, London, 1993.

28 A.N.E. Birch et al, 'Interactions between plant resistance genes, pest aphid populations and beneficial aphid predators', Scottish Crop Research Institute, Dundee, Annual Report 1996/97 pp.68-72.

29 'Buildup of Bt toxins in soil', The Gene Exchange—A Public Voice on Biotechnology and Agriculture, Union of Concerned Scientists, Fall/Winter 1998 <www.ucsusa.org/publications/index.html>; C. Crecchio and G. Stotzky, 'Insecticidal activity and biodegradation of the toxin from Bacillus thuringiensis subsp. kurstaki bound to humic acids from soil', Soil Biology and Biochemistry, Vol.30, pp.463-70, 1998.

30 Prof. J. Cummins, 'The Danger of Virus-Resistant Crops' <www.psrast.org/ctenvir.htm>.

31 James Kling, 'Could Transgenic Supercrops One Day Breed Superweeds?', Science Vol.274, October 1996; M.G. Paoletti and D. Pimentel, 'Genetic Engineering in Agriculture and the Environment: assessing risks and benefits', BioScience, Vol.46, 1996, pp.665-671; R.A. Steinbrecher, 'From Green to Gene Revolution: the environmental risks of genetically engineered crops', The Ecologist, Vol.26, 1996, pp.273-282.

32 Kurt Kleiner, 'Fields of Genes', New Scientist, 16 August 1997, p.4.

33 As 33.

34 J. S. Cory, 'Release of Genetically Modified Viruses", Reviews in Medical Virology, Vol.1, 1991, pp.79-88.

35 Jeremy Rifkin, The Biotech Century, Tarcher and Putnam, New York, 1998, p.67.

36 John Clamp, 'Genetic Maize Growing Will Pollute The Environment', Totnes Times, Totnes, Devon, 23 April 1998.

37 Julie Shepherd, 'From BSE to Genetically Modified Organisms: Science, Uncertainty and the Precautionary Principle', Briefing prepared for Greenpeace UK, July 1997, pp.28-29.

38 'The Precautionary Principle' in Rachel's Environment & Health Weekly No.586, 19 February 1998 <www.psrast.org/precaut.htm>.

39 René von Schomberg, 'An appraisal of the working in practice of directive 90/220/EEC on the deliberate release of genetically modified organisms', Scientific and Technological Options Assessment (STOA) of the European Parliament, 2 January 1998 <www.europarl.eu.int/dg4/stoa/en>.

40 D. M. Conning, 'Biotechnology—Influencing Public Opinion' in BCPC Monograph No.55: Opportunities for molecular biology in crop protection, pp.299-304.

41 Thomas Epprecht, 'Genetic engineering and liability insurance—the power of public perception', Swiss Reinsurance Company, 1998.

42 Greg and Pat Williams, 'Risk of Genes Escaping from Transgenic Crops', Hortideas, Vol.14, No.4, April 1994.

43 J.F. Hancock, R. Grumet and S. Chokanson, 'The Opportunity for Escape of Engineered Genes from Transgenic Crops', HortScience Vol.31 No.7, December 1997, pp.1080-1085.

44 S. Frello, K.R. Hansen, J. Jensen J and R.B. Joergensen, 'Inheritance of Rapeseed (Brassica napus) Specific RAPD Markers and a Transgene in the Cross B. juncea x (B. juncea x B. napus)', Theor. Appl. Genet. Vol.91, 1995, pp.236-241; R.B. Joergensen and B. Andersen (1994), 'Spontaneous Hybridization Between Oilseed Rape (Brassica napus) and Weedy B. campestris (Brassicaceae): a Risk of Growing Genetically Modified Oilseed Rape', Am.J.Botany, Vol.81, pp.1620-1626; T.R. Mikkelsen, B. Andersen and R.B. Joergensen, 'The Risk of Crop Transgene Spread', Nature, Vol.380, 1996, p.31.

45 'Genetically Engineered Oilseed Rape: Agricultural Saviour or New Form of Pollution?' GeneWatch Briefing No.2, May 1998; Agrow, 296, 16 January 1998, p.9.

46 'Update on Risk Research—More on Transgenes in Wild Populations', *The Gene Exchange—A Public Voice on Biotechnology and Agriculture*, Union of Concerned Scientists, Fall/Winter 1998 <www.ucsusa.org/publications/index.html>; A. Snow and R. Jorgensen, 'Costs of transgenic glufosinate resistance introgressed from *Brassica napus* into weedy *Brassica rapa*', Abstract of a paper presented at the annual meeting of the Ecological Society of America, Baltimore, MD, 6 August 1998; A. Dove et al., 'Research news: promiscuous pollination', *Nature Biotechnology*, Vol.16, p.805, September 1998.

47 'Update on Risk Research—Process Counts', *The Gene Exchange—A Public Voice on Biotechnology and Agriculture*, Union of Concerned Scientists, Fall/Winter 1998. <www.ucsusa.org/publications/index.html>; J. Burgelson, C.B. Purrington and G. Wichmann, 'Promiscuity in transgenic plants', *Nature*, Vol.395, p.25, 3 Sept 1998.

48 L. Spinney, 'Biotechnology in Crops: Issues for the developing world',A report compiled for Oxfam Great Britain, May 1998 <www.oxfam.org.uk/policy/papers/gmfoods/gmfoods.htm>.

49 R. Steinbrecher and M. Ho, 'Fatal Flaws in Food Safety Assessment: Critique of the joint FAO/WHO Biotechnology and Food Safety Report', 3.2, 1996; P. J. Regal, 'Scientific principles for ecologically based risk assessment of transgenic organisms', *Molecular Ecology*, Vol.3, 1994, pp.5-13 <www.psrast.org/pjrisk.htm>.

50 'Joint FAO/WHO Expert Consultation on Biotechnology and Food Safety', Rome, 1996, p.20 <www.fao.org/waicent/faoinfo/economic/esn/biotech/tabconts.htm>.

51 Taken from a series of quotes on the Natural Law Party Wessex web pages. <www.btinternet.com/~nlpwessex/Documents/contentsfall.htm>

52 Allison Snow and Pedro Morán Palma, 'Commercialisation of Transgenic Plants: Potential Ecological Risks', *Bioscience*, February 1997, p.94; Jeremy Rifkin, *The Biotech Century*, Tarcher and Putnam, New York, 1998, p.77

53 'Genetic Genie: The Premature Commercial Release of Genetically Engineered Bacteria', Public Employees for Environmental Responsibility, 1995. Cited in Brian Tokar, 'What is Biotechnology?', Fact Sheet on Genetically Engineered Foods & Crops <www.purefood.org/ge/geFactSheet.htm>.

54 'Field trial of a transgenic arthropod, *Metaseilulus occidentalis* (Acari: Phytoseiidae)', APHIS (1996), Field Trial Report; G. Naik (1997), 'Turning mosquitoes into malaria fighters', *Dow Jones News*, 17 June 1997; 'Field trial of a transgenic nematode, *Heterorhabditis bacteriophora* (Nematoda: Heterorhabditidae)', APHIS Field trial report, 1996.

55 Stephen Nottingham, *Eat Your Genes*, Zed Books, London, 1998, pp.80-82.

56 'The Case of the Competitive *Rhizobia*', US National Biotechnology Impacts Assessment Programme Newsletter, March 1991.

57 F. Gebhard and K. Smalla, 'Transformation of *Acinetobacter* sp. strain BD413 by transgenic sugar beet DNA', *Appl. Environ. Microbiol.*, Vol.64, 1998, pp.1550-1559.

58. F. Gebhard and K. Smalla, 'Monitoring field releases of genetically modified sugar beets for persistence of transgenic plant DNA and horizontal gene transfer', *Microbiology Ecology* 28, 1999, pp.261-272.

59 M.T. Holmes et al., 'Effects of *Klebsiella planticola* on soil biota and wheat growth in sandy soil', *Applied Soil Ecology*, Vol.326, 1998, pp.1-12; Union of Concerned Scientists, *The Gene Exchange—A Public Voice on Biotechnology and Agriculture*, Fall/Winter 1998 <www.ucsusa.org/publications/index.html>; Personal communication with Michael Holmes and Elaine Ingham, April 1999.

60 'Fishy business?', *The Splice of Life*, Vol.3 No.3, December 1996 <www.geneticsforum.org.uk/>.

61 'Super Fish', *Equinox*, March/April 1987; Mike Toner, 'Cultivating Designer Fish',

Atlanta Journal, 21 May 1991.

62 Shao Jun Du et al, 'Growth Enhancement in Transgenic Atlantic Salmon by the Use of an "All Fish" Chimeric Growth Hormone Gene Construct', *Bio Technology* Vol.10 No.2, 1992, pp.176-181.

63 Steven Nottingham, *Eat Your Genes*, Zed Books, London, 1998, p.88.

64 D. MacKenzie, 'Can we make supersalmon safe?', *New Scientist*, 27 January 1996, pp.14-15; 'New Prospects for Gene Altered Fish Raise Hope and Alarm', *New York Times*, 27 November 1990.

65 'Fletcher Challenge Forests, International Paper, Monsanto Company and Westvaco Corporation Announce Forestry Biotechnology Joint Venture', Company Press Release, New York, 6 April 1999.

66 Jeremy Rifkin, *The Biotech Century*, Tarcher and Putnam, New York, 1998, p.74.

67 Brian Tokar, 'What is Biotechnology?', Fact Sheet on Genetically Engineered Foods & Crops <www.purefood.org/ge/geFactSheet.htm>.

68 Personal communication with Christine Von Weiszacker, April 27,1999 and reference to I.Kowarik, 'Ecological Consequences of the Introduction and Dissemination of New Plant Species: An analogy of the release with genetically engineered organisms', *BBL*, Bonn, 1990.

69 'Green Group Warns on GM Tree Development', Reuters, Brussels, Tuesday 9 November 1999.

70 C. James, 'Global Review of Commercialised Transgenic Crops: 1998', ISAAA Briefs No.8, ISAAA: Ithaca, NY.

Chapter 3 GE and Farming

1 Cornish Farmer Michael Hart during his tour of the South-West with the 'Keep Britain Farming' roadshow. Reported in *Blackmore Vale Magazine*, Dorset, 14 August 1998.

2 A. Johnston (1989), 'Biological nitrogen fixation' in *A Revolution in Biotechnology* (ed. J.L. Marx), Cambridge University Press, Cambridge/ New York, pp.103-118.

3 M.-W. Ho (1998), 'Genetic Engineering: Dream or Nightmare?', Gateway Books, Bath, UK, p.135.

4 Union of Concerned Scientists, *The Gene Exchange—A Public Voice on Biotechnology and Agriculture*, Fall 1997 <www.ucsusa.org/Gene/F97.glyphosate.html>; 'Monsanto Checks Cotton Problems', *Commercial Appeal*, Memphis, Tennessee, 16 August 1997 (via *Bloomberg News online*); 'Mississippi Investigating Monsanto's Cotton', *Commercial Appeal*, Memphis, Tennesse, 16 August 1997 (via *Bloomberg News online*); B. Reid, 'Problems Crop Up with New Cotton Variety', *Clarion-Ledger*, Jackson, Mississippi, 20 August 1997; B. Reid, 'Genetic Cotton Backfires', *Clarion-Ledger*, Jackson, Mississippi, 14 September 1997; B. Reid, 'New Breed of Cotton Raises More Questions', *Clarion-Ledger*, Jackson, Mississippi, 24 September 1997.

5 Union of Concerned Scientists, *The Gene Exchange—A Public Voice on Biotechnology and Agriculture*, Summer 1998 <www.ucsusa.org/Gene/su98.arbitrate.html>; Mississippi Department of Agriculture and Commerce, Seed Arbitration Council, 'Recommendation of settlement: Re-Thom Farms, Romar Farms, and Talley Planting Co. v. Delta and Pine Land, Monsanto, and Paymaster Technology', 6 December 1998.

6 Allen R. Myerson, 'Seeds of Discontent: Cotton Growers Say Strain Cuts Yields', *New York Times*, 19 November 1997.

7 L. Spinney, 'Biotechnology in Crops: Issues for the developing world', A report compiled for Oxfam Great Britain, May 1998 <www.oxfam.org.uk/policy/papers/gmfoods/gmfoods.htm>; 'Monsanto's transgenic potatoes on the loose in Georgia (1996-1998): the need for an international Biosafety Protocol', Greenpeace International, Amsterdam, 1998.

8 Union of Concerned Scientists, 'Post-Approval Blues: FlavrSavr Tomato—Squashed', *The Gene Exchange—A Public Voice on Biotechnology and Agriculture*, Fall 1997 <www.ucsusa.org/Gene/F97.agribusiness.html#blues>; 'News Release—Calgene Announces Second Quarter Financial Results', Calgene, 6 February 1996; J. Bleifuss, 'Recipe For Disaster', *In These Times*, Chicago, Illinois, 11 November 1996 <www.purefood.org/recipe.html>; R. King, 'Low-Tech Woe Slows Calgene's Super Tomato', *Wall Street Journal*, 11 April 1996, p.B1; 'The Cutting Edge', *Los Angeles Times*, 18 August 1997.

9 'Bt Cotton Fails to Control Bollworm', *The Gene Exchange—A Public Voice on Biotechnology and* Agriculture, Union of Concerned Scientists, Winter 1996 <www.ucsusa.org/Gene/W96.bt.html>; PANUPS (Pesticide Action Network North America Updates Service), 9 December 1996 <www.rtk.net/E15969T598>; M. Woodfin, 'Bt cotton creating resistance to Bt?', *Southern Sustainable Farming* No.12, September 1996 <www.pmac.net/bt2.htm>; W. Board, 'Bt cottons not immune to injury despite benefits' <www.lubbockonline.com/news/111596/bt.htm>; 'Plant smart to avoid Bt-resistant corn borer, experts advise', *Purdue News*, October 1996 <www.purdue.edu/UNS/html4ever/9610.Bledsoe.BtCorn.html>; R. Nigh, 'Bt cotton', 13 November 1996 <http://metalab.unc.edu/london/permaculture/mailarchives/sanet1/0651.html>; C. Hagedorn, 'The Bollworm Controversy—Monsanto's Bt Cotton in 1996', *Crop and Soil Environmental News*, January 1997 <www.ext.vt.edu/news/periodicals/cses/1997-01/1997-01-01.html>.

10 David Holzman, 'Agricultural Biotechnology: Report Leads to Debate on Benefits of Transgenic Corn and Soybean Crops', *Genetic Engineering News*, Vol.19,No.8, 15 April 1999.

11 Marc Lappé and Britt Bailey, *Against the Grain*, Earthscan, London, 1999, pp.82-3.

12 As 11, p.59.

13 Bill Christison, president of US National Family Farm Coalition, speaking at Chillicote, Missouri, at the First Grassroots Gathering on Biodevastation: Genetic Engineering. St Louis, Missouri 18 July 1998: taken from a series of quotes on 'Will GM crops deliver benefits to farmers?' on the web page <www.btinternet.com/~nlpwessex/Documents/contentsfall.htm>.

14 J. Berlan and C. Lewontin, 'Cashing in on Life—Operation Terminator', *Le Monde Diplomatique*, December 1998 <www.monde-diplomatique.fr/en/1998/12/02gen.html>.

15 Kenny Bruno, 'Say It Ain't Soy, Monsanto', *Multinational Monitor*, Volume 18 Numbers 1 and 2, January/February 1997 <www.purefood.org/aintsoy.html>.

16 Richard Ford, 'What's law on GMO's?', Farmers Weekly (UK), 19 March 1999.

17 As reported in *The Washington Post* of 3 February 1999 and *The International Herald Tribune* of 4 February 1999, 'Monsanto Sues North American Farmers', Friends of the Earth Biotech Mailout, Volume 5, Issue 2, 15 March 1999. <www.foeeurope.org/programmes/biotechnology/5n2_frames.htm>.

18 As 17.

19 Quote from Percy Schmeiser in *Western Producer*, November 1998. Taken from a series of quotes 'Will GM crops deliver benefits to farmers?' on the Natural Law Party Wessex webpage <www.btinternet.com/~nlpwessex/Documents/contentsfall.htm>.

20 Quote from Nick Brown on 'The Jonathan Dimbleby programme', ITV, London, 31 January 1999.

21 Minutes from ACRE meeting June 1998, item 1.1: 'GM maize in National List trials adjacent to an organic farm in Devon', ACRE\98\P25, UK.

22 J. Emberlin, 'The Dispersal of Maize Pollen', National Pollen Research Unit, 2 March 1999. Copies available from The Soil Association, Bristol.

23 Nick Nuttall, 'Bees spread genes from GM crops', *The Times*, London, 15 April 1999.

24 L. Keenan, 'First Case of GMO Food Contamination', Genetics Food Alert Press Release, 4 February 1999 <www.essential-trading.co.uk/genetix.htm>.

25 R. Cummins and B. Lilliston, *In Motion* magazine, Little Marais, Minnesota <www.inmotionmagazine.com/geff.html>.

26 C. Cairns, 'GM Farmers have Grounds for Concern—Experts warn of fall in land values similar to effect of contamination or disease', *The Scotsman*, 11 March 1999.

27 *Science*, Vol.272, 11 October 1996, pp.180-181. Cited in Julie Shepherd, 'From BSE to Genetically Modified Organisms: Science, Uncertainty and the Precautionary Principle', Briefing prepared for Greenpeace UK, July 1997, p.26.

28 'President Clinton Expands Federal Effort to Combat Invasive Species', USDA Press Release, 3 February 1999 <www.doi.gov/news/990203.html>.

29 D. Pimentel, L. Lach, R. Zuniga and D. Morrison, *Environmental and Economic Costs Associated with Non-indigenous Species in the United States*, Cornell University, College of Agriculture and Life Sciences <www.news.cornell.edu/releases/Jan99/species_costs.html>; Alan Hall, 'Costly Interlopers—Introduced species of animals, plants and microbes cost the U.S. $123 billion a year' <www.sciam.com/explorations/1999/021599animals/index.html>.

30 R.A. Malecki, B. Blossy, S.D. Hight, D. Schroeder, L.T. Kok and J.R. Coulson, 'Biological control of purple loosestrife', *BioScience*, Vol.43 No.10, 1993, pp.680-686.

31 D.G. Thompson, R.L. Stuckey and E.B. Thompson, 'Spread, impact, and control of purple loosestrife (*Lythrum salicaria*) in North American wetland', *Fish and Wildlife Research* 2, US Fish and Wildlife Service, Washington, DC, 1987.

32 ATTRA, 'Purple Loosestrife: Public Enemy #1 on Federal Lands', ATTRA, Washington, DC, 1997 <http//refuges.fws.gov/NWRSFiles/HabitatMgmt/PestMgmt/LoosestrifeProblem.html>.

33 Bernard Rollin, *The Frankenstein Syndrome: Ethical and Social Issues in the Genetic Engineering of Animals*, Cambridge University Press, New York, 1995, p.119.

34 N. Myers, 'Biodiversity and the precautionary principle', *Ambio*, Vol.22(2-3), 1993, pp.74-79; E.O. Wilson, *The Diversity of Life*, Penguin Books, 1994, p.268.

35 'Crop Genetic Resources' in *Biodiversity for food and agriculture*, FAO, Rome, 1998.

36 M. Altieri, 'The Environmental Risks of Transgenic Crops: an Agroecological Assessment', Department of Environmental Science, Policy and Management, University of California, Berkeley <www.pmac.net/miguel.htm>.

37 N. Alexandratos, 'World Agriculture: Toward 2000', an FAO Study, FAO/ Belhaven, Rome and London, 1988.

38 'Genetic Engineering and World Hunger', Corner House Briefing No.10, October 1998.

39 S. Brush, 'Farming on the edge of the Andes', *Natural History* (5) 1977, pp.32-41; Marc Lappé and Britt Bailey, *Against the Grain*, Earthscan, London, 1999, p.99.

40 'The potato blight is back' in *Seedling*, the quarterly magazine of GRAIN (Genetic Resources Action International) <www.grain.org/publications/oct95/oct952.htm>.

41 'Greed or need? Genetically modified crops', Panos Media Briefing No.30, 1998.

42 V. Shiva, 'Monocultures, Monopolies, Myths and the Masculinisation of

Agriculture', *Aisling Quarterly*.

43 As 42.

44 Laura Spinney, 'Biotechnology in Crops: Issues for the developing world', Research paper compiled for Oxfam GB, May 1998 <www.oxfam.org.uk/policy/papers/gmfoods/gmfoods.htm>.

45 'Genetic Engineering and World Hunger', Corner House Briefing 10, October 1998; 'Agricultural research for whom?', an article edited by *The Ecologist* from research material provided by GRAIN and RAFI, *The Ecologist*, Vol.26, No.6, November/December 1996.

46 'Genetic Engineering and World Hunger', Corner House Briefing 10, October 1998.

47 As 44.

48 'IPM-trained farmers in Indonesia escape pest outbreaks' <www.fao.org/NEWS/1998/981104-e.htm>.

49 Marc Lappé and Britt Bailey, *Against the Grain*, Earthscan, London, 1999, p.102.

50 WHO estimates in J. Jeyaratnam, 'Acute Pesticide Poisoning: A Major Global Health Problem,' *World Health Statistics Quarterly*, 1990, pp.139-144

51 J. Jeyaratnam, 'Health Problems of Pesticide Usage in the Third World,' *British Journal of Industrial Medicine*, Vol.42, 1995, pp.505-506.

52 'UN Convention to regulate trade in hazardous pesticides', FAO press release <www.fao.org/news/1997/971108%2De.htm>; 'Convention on Trade in Dangerous Chemicals and Pesticides: Round of Workshops Started', FAO Press Release, Bangkok, 8 December <www.fao.org/ag/agp/agpp/pesticid/pic/picnews8.htm>.

53 Lester Brown et al, *State of the World 1997*, Worldwatch Institute Report, 1997, p.16.

54 'African Scientists Condemn Advertisement Campaign for Genetically Engineered Food: Call for European Support', Gaia Foundation Press Release, 3 August 1998 <www.psrast.org/afrscimo.htm>.

55 Quoted in J. Pretty, 'Feeding the world with sustainable farming or GMOs?', *Splice*, the magazine of The Genetics Forum (UK), Vol.4, Issue 6, Aug/Sept 1998, pp.4-5.

56 'Global Development Finance 1999—Annual and Summary Tables', The World Bank Group, Washington DC, 30 March 1999, p.160 <www.worldbank.org/prospects/gdf99/vol1.htm>; 'Official Development Flows in 1997', OECD, Paris, February 1999.

57 As 46.

58 Stephen Nottingham, *Eat Your Genes*, Zed Books, London, 1998, p.157.

59 Frances Moore Lappé, Joseph Collins and Peter Rosset, *World Hunger: Twelve myths*, Grove Press, New York, 1998.

60 Lester Brown et al, *State of the World 1997*, Worldwatch Institute Report, 1997, p.47.

61 As 60, p.49. <www.oxfam.org.uk/policy/papers/gmfoods/gmfoods.htm>.

62 Miguel Altieri, Agroecology: the science of sustainable agriculture, Boulder, Colorado: Westview Press, 1995; Peter Rosset, 'The Crisis of Modern Agriculture: Toward an Agroecological Alternative', Paper presented at the fourth PAN International meeting 'Feeding the World Without Poisons: Supporting Healthy Agriculture', Santa Clara, Cuba, May 17-21, 1997, p.14.

63 Peter Rosset, 'The Crisis of Modern Agriculture: Toward an Agroecological Alternative', Paper presented at the fourth PAN International meeting 'Feeding the World Without Poisons: Supporting Healthy Agriculture', Santa Clara, Cuba, 17-21 May 1997, p.15.

64 'Farming Systems Trial', Rodale Institute, Kutztown, PA, USA, 1981-1995 <www.enviroweb.org/publications/rodale/usrarc/fst.html>.

65 Quoted in J. Pretty, 'Feeding the world with sustainable farming or GMOs?', *Splice*, the magazine of The Genetics Forum (UK), Vol.4, Issue 6, Aug/Sept 1998, pp.4-5.

66 US National Research Council, *Lost Crops of Africa: Volume 1, Grains*, cited in Laura Spinney, 'Biotechnology in Crops: Issues for the developing world', research paper compiled for Oxfam GB, May 1998

67 'World Census on Agriculture', FAO Census Bulletins, Rome, 1980.

68 As 46.

69 As 46.

70 Lester Brown et al, *State of the World 1997*, Worldwatch Institute Report, 1997, p.40.

71 Marc Lappé and Britt Bailey, *Against the Grain*, Earthscan, London, 1999; Marc Lappé, *Diet for a Small Planet*, Ballantine, New York, quoted in J.Doyle, *Altered Harvest*, Viking Penguin, New York, 1985, p.287; Jeremy Rifkin, *Beyond Beef: The Rise and Fall of the Cattle Culture*, Penguin, New York, 1992, p.160-161.

72 W. Bello and S. Rosenfeld, 'Dragons in Distress: Asia's Miracle Economies in Crisis', Institute for Food and Development Policy, San Francisco, 1990, p.86.

73 'The Sustainable CEO: Monsanto', *Impact*, Vol.2 No.2, Spring 98.

74 'Monsanto 1997 Report on Sustainable Development', Monsanto Corporation, St. Louis, Missouri, p.27.

N. Dawkins and D. Wysham, 'The World Bank's Consultative Group to Assist the Poorest: Opportunity or Liability for the World's Poorest Women?', Institute for Policy Studies, Washington DC, USA.

75 Laura Spinney, 'Biotechnology in Crops: Issues for the developing world', research paper compiled for Oxfam GB, May 1998
<www.oxfam.org.uk/policy/papers/gmfoods/gmfoods.htm>

76 V. Shiva, 'Betting on Biodiversity: Why Genetic Engineering Will Not Feed the Hungry', Research Foundation for Science, Technology and Ecology, India, 1998, p.36.

77 L. Busch et al, *Plants, Power and Profit*, Basil Blackwell, Oxford, 1990.

Miguel Altieri, 'Biotechnology Myths'. <www.purefood.org/ge/biomyth.html>

78 As 75 and 76.

79 Margaret Mellon, 'Dead Seeds', *The Gene Exchange—A Public Voice on Biotechnology and Agriculture*, Union of Concerned Scientists, Fall/Winter 1998
<www.ucsusa.org/publications/index.html>.

80 Hope Shand and Pat Mooney, 'Terminator Seeds Threaten an End to Farming', *Earth Island Journal*, Fall 1998
<www.earthisland.org/eijournal/fall98/fe_fall98terminator.html>.

81 'The Debate on Genetically Modified Organisms: Relevance for the South', Overseas Development Institute Briefing Paper, January 1999
<www.oneworld.org/odi/briefing/1_99.html>.

82 As 80.

83 As 80.

84 'Terminator Technology and the Developing World', *The Gene Exchange—A Public Voice on Biotechnology and Agriculture*, Union of Concerned Scientists, Fall/Winter 1998 <www.ucsusa.org/publications/index.html>.

85 'Terminator Terminated?', RAFI News Release, Rural Advancement Foundation International, 4 October 1999; 'RAFI on Monsanto merger: Pharma-geddon', 'Geno-Types', Rural Advancement Foundation International, 21 December 1999 <www.rafi.org>; Philip Brasher, 'Terminator Seeds', Associated Press, Washington, 31 October 1999.

86 Pat Mooney, quoted in 'Genetic Seed Sterilisation is "Holy Grail" for Agricultural Biotechnology Firms—New Patents for "Suicide Seeds" Threaten Farmers and Food

Security Warns RAFI', RAFI press release <www.rafi.org/pr/release26.html>.
87 'Traitor Technology—Damaged Goods from the Gene Giants', RAFI News Release, 29 March 1999 <www.rafi.org>.

Chapter 4 Patenting Life

1 Ted Howard, 'The Case Against Patenting Life', brief on behalf of the People's Business Commission, Amicus Curiae, in the Supreme Court of the United States No.79-136, 29.

2 'New Developments in Biotechnology: Patenting life—Special Report', US Congress, Office of Technology Assessment, OTA-BA-370, US Government Printing Office, April 1989, p.37.

3 Sidney A. Diamond, Commissioner of Plants and Trademarks, petitioner, v. Ananda M. Chakrabarty et al., 65 L ed. 2d 144, 16 June 1980, p.152.

4 Giovanna Brel, 'An Illinois Biochemist Wins a Crucial Patent Fight, and a New Era of Life in a Test Tube Begins', *People*, 14 July 1980, p.38.

5 Quoted in Brian Belcher and Geoffrey Hawtin, *A Patent on Life: Ownership of Plant and Animal Research*, IDRC, Canada, 1991.

6 Jeremy Rifkin, *The Biotech Century*, Tarcher and Putnam, 1998, p.43.

7 H. Hobbelink, *Biotechnology and the Future of World Agriculture*, Zed Books, London, 1991.

8 'Patenting, Piracy and Perverted Promises', briefing published by Genetic Resources Action International (GRAIN), Barcelona, Spain, 1998.

9 W. Lambert and A.S. Hayes, 'Investing in patents to file suits is curbed', *Wall Street Journal*, 30 May 1990.

10 'Patenting, Piracy and Perverted Promises', briefing published by Genetic Resources Action International (GRAIN), Barcelona, Spain, 1998; 'Bioprospecting /Biopiracy and Indigenous Peoples', *RAFI Communiqué*, published by the Rural Advancement Foundation International; F. Powledge, 'Who Owns Rice and Beans?' *BioScience*, July/August 1995 pp.440-444.

11 As 8; 'Bioprospecting/Biopiracy and Indigenous Peoples', *RAFI Communiqué*, published by the Rural Advancement Foundation International.

12 Vandana Shiva, *Biopiracy: The Plunder of Nature and Knowledge*, Green Books, 1998, p.58.

13 Quoted in Hope Shand, 'Patenting the Planet', *Multinational Monitor*, June 1994, p.13. Also 'Species patent on transgenic soybeans granted to transnational chemical giant W.R. Grace', *RAFI Communiqué*, 1994 <www.rafi.org/communique/19942.html>.

14 European Patent Office, 'Method for producing transgenic animals', Harvard College, European Patent No.0169 672; L. Raines, 'Of Mice and Men and Tennis Balls', *Across the Board*, March 1989, p.46.

15 Andrew Kimbrell, *The Human Body Shop*, Regnery Publishing, Washington DC, 1997, p.236.

16 'New Developments in Biotechnology: Patenting life—Special Report', US Congress, Office of Technology Assessment, OTA-BA-370, US Government Printing Office, April 1989, p.37.

17 Testimony of Michael Glough, US Congress, Office of Technology Assessment, Before the Subcommittee on Intellectual Property and Judicial Administration House Committee on the Judiciary, 20 November 1991, on 'Patents and Biotechnology'; As 15, pp.237-238; Patenting data compiled by the International Center for

Technology Assessment, Washington DC.

18 As 8.

19 A. Coghlan, 'One small step for a sheep', *New Scientist*, 1 March 1997.

20 Jeremy Rifkin, *The Biotech Century*, Tarcher and Putnam, 1998, p.47.

21 Correspondence from Ma Yong-woon, Biosafety Campaigner at the Environmental Policy Department, Korean Federation for Environmental Movement (KFEM). Reported in the South Korean Newspaper *Chosun-Ilbo* on 14 December 1998.

22 C. Cookson, 'Cloned body parts, not babies, may get the thumbs up', *Financial Times*, London, 9 December 1998; The Human Genetic Advisory Commission and the Human Fertilisation and Embryology Authority, 'Cloning issues in reproduction science and medicine', Dept. of Trade and Industry, London, 1998.

23 Stephen Nottingham, *Eat Your Genes*, Zed Books, London, 1998, p.33.

24 P. Dixon, *The Genetic Revolution*, Kingsway, Eastbourne, 1995.

25 'Flourescent mice—a green light for farm animal experiments', *Agscene*, Autumn 1997, p.18.

26 Brian Tokar, 'Send in the Clones?', *Food & Water Journal*, Spring 1997. Harvey Bialy, 'Barnyard Biotechnology:Transgenic Pharming Comes of Age', *Biotechnology*, Vol.9, September 1991.

27 Tim O'Brien, 'Genetic Engineering', *Compassion in World Farming Fact Sheet*, October 195, p.3.

28 As 15, pp.221-3.

29 'Genetic Engineering?' *Agscene* No. 103, Summer 1991, p.22. *New Scientist*, 14 November 1992, pp.13-14.

30 C. Tudge, *The Engineer in the Garden. Genetics: From the Idea of Heredity to the Creation of Life*, Jonathan Cape, London, 1993.

31 Stephen Nottingham, *Eat Your Genes*, Zed Books, London, 1998, p.99.

32 V.G. Pursel et al, 'Genetic Engineering of Livestock', *Science*, Vol.244, 1989, pp.1281-88.

33 R.E. Hammer et al, 'Production of transgenic rabbits, sheep and pigs by microinjection', *Nature*, Vol.315, 1985, pp.680-683; C.E. Rexroad et al, 'Production of transgenic sheep with growth-regulating genes', *Molecular Reproduction and Development*, Vol.1, 1989, pp.164-169; C.E. Rexroad et al, 'Transferrin-directed and albumin-directed expression of growth-related peptides in transgenic sheep', *Journal of Animal Science*, Vol.69, 1991, pp.2995-3004.

34 US Patent and Trademark Office, 'Animals–Patentability', Washington DC: US Patent and Trademark Office, 7 April 1987.

35 As 28, p.239.

36 Opinion, 'Moore v. The regents of the University of California et al', Supreme Court of the State of California, file No.S006987, slip op. 2, 3: William Carlsen, 'Key ruling by State Court On Body Cells', *San Francisco Chronicle*, 10 July 1990.

37 As 8 and 15.

38 As 8.

39 Moore v. The Regents of the University of California et al, Supreme Court of the State of California, 16.

40 As 8.

41 'Study Finds Public Science Is Pillar Of Industry', *The New York Times*, May 13, 1997, Section C, p.1

42 Julian Borger, 'Rush to patent genes stalls cures for disease', *The Guardian*, December 15th, 1999
<www.guardianunlimited.co.uk/Archive/Article/0,4273,3941983,00.html>

43 'No Patents on Life!', The Council for Responsible Genetics <www.essential.org/crg/nopatents.html>.

44 'Patenting, Piracy and Perverted Promises', briefing published by Genetic Resources Action International (GRAIN), Barcelona, Spain, 1998; 'The Debate on Genetically Modified Organisms: Relevance for the South', Overseas Development Institute Briefing Paper, 1 January 1999 <www.oneworld.org/odi/briefing/1_99.html>; 'Genetic Patenting', Hansard (House of Commons Daily Debates), 28 July 1997, 10.13pm <www.parliament.the-stationery-office.co.uk/pa/cm199798/cmhansrd/ cm970728/debtext/70728-35.htm>; Rita Rubin, 'A hard sell to bank your baby's blood', US News Online <www.usnews.com/usnews/issue/29cord.htm>; 'European Patent on Baby Blood Revoked!', European Campaign On Biotechnology Patents Press Release, Munich, 8 June 1999.

45 As 28, p.252.

46 United States Patent No.5,061,620, 29 October 1991, 'Human Hematopoetic Stem Cell'.

47 'Gene Boutiques Stake Claim to Human Genome', RAFI Communiqué, May/June 1994 <www.rafi.org/communique/19943.html>.

48 'Corporations receive patents on human genes and their products', RAFI Communiqué, May/June 1995 <www.rafi.org/communique/fltxt/19953.html>.

49 Usha Lee McFarling, 'The code war: Biotech firms engage in high-stakes fight over rights to the human blueprint', San José Mercury News, 17 November 1998 <www.mercurycenter.com/premium/scitech/docs/rgenome17.htm>.

50 'The Gene Giants: Masters of the Universe?', RAFI Communiqué, March/April 1999 <www.rafi.org/communique/fltxt/19992.html>.

51 'The Human tissue trade', RAFI Communiqué, Jan/Feb 1997.

52 As 8.

53 'The Human Genome Diversity Project', Indigenous People Coalition Against Biopiracy, 1998 <www.niec.net/ipcb/research/thehgdp.html>; K.Y. Kreeger, 'Proposed Human Genome Diversity Project still plagued by controversy and questions', The Scientist, Vol.10 No.20, 1996, pp.1-8.

54 C. Juma, The Gene Hunters: Biotechnology and the Scramble for Seeds, Zed Books, London, 1989.

55 As 12, pp.73-75; Michael Hirsh, 'Fight For the Miracle Tree', Bulletin, 26 September 1995, pp.70-71; K.Vijayalakshmi et al (1995), Neem: A User's Manual, Centre for Indian Knowledge Systems and Research Foundation for Science, Technology and Natural Resource Policy, New Delhi.

56 As 12, p.73.

57 As 8.

58 'Novel Pharmaceuticals from Genetically Engineered Fungi', Myco Pharmaceuticals, Corporate Profile, 9 March 1994, p.3.

59 Kathy Heine, 'Treasure in the Jungle', Monsanto Magazine, April 1991, No.1, p.22.

60 Franklin Hoke, 'Recently Ousted Genentech President G. Kirk Raab Named Board Chairman At Shaman Pharmaceuticals', The Scientist, Vol.9 No.20, 16 October 1995, p.17.

61 J. Madeley, Yours For Food, Christian Aid, UK, 1996.

62 'Biodiversity: Protections Provided in International Pacts Seen As Best Framework for Bioprospecting', BNA International Environment Daily, May 1993; Elissa Blum, 'Making Biodiversity Conservation Profitable: A Case Study of the Merck/INBio Agreement', Environment Vol.35 No.4, May 1993, pp.16-20 & pp.38-44; 'Convention

on Biological Diversity', United Nations Environment Programme, United Nations Doc. Na.92-7807: 9-12; Richard Gardner, *Negotiating Survival: Four Priorities After Rio*, Council on Foreign Relations Press, New York, 1992; *GATT Focus 98* (April 1993), pp.1-4; 'If Biological Diversity Has A Price, Who Sets It and Who Should Benefit?', *Nature* 359, 15 October 1992; Christopher Joyce, 'Western Medicine Men Return to the Field: Tropical Forest Loss and Fast Lab Techniques are Propelling the Search for Therapeutic Phytochemicals', *BioScience* Vol.42 No.6, June 1992, pp.399-403; Jeffery McNeely et al, *Conserving the World's Biological Diversity*, The World Bank, World Resources Institute, International Union for Conservation of Nature and Natural Resources, Conservation International and the World Wildlife Fund: Gland, Switzerland and Washington, DC, 1990; 'New Measure Would Cover Extraction of Genetic Resources from Rain Forest', *BNA International Environmental Daily*, 21 July 1992; Leslie Roberts, 'The Merck Story: Serving Society', Public Affairs Department, Merck & Company, May 1993; 'Chemical Prospecting: Hope for Vanishing Ecosystems?', *Science* 256, 22 May 1992, pp.1142-3; Report from the Instituto Nacional de Biodiversidad (INBio), December 1992; Nathaniel Sheppard Jr, 'Costa Rica Makes Conservation Pay Country to Benefit From Research Using Its Resources', *Chicago Tribune*, 12 August 1993; 'The Biodiversity Treaty; Pandora's Box or Fair Deal?', *Science* 256, June 19, 1992, p.1624; Sam Thernstrom, 'Jungle Fever: Lost Wonder Drugs of the Rainforest', *The New Republic* 208, 19 April 1993.

63 'Bioprospecting/Biopiracy And Indigenous Peoples', *RAFI Communiqué*, published by the Rural Advancement Foundation International.

64 Jeremy Rifkin, *The Biotech Century*, Tarcher and Putnam, 1998, p.54.

65 J. Enyart, 'A GATT Intellectual Property Code', *Les Nouvelles*, June 1990, pp.54-56.

66 Leon R. Kass, 'Patenting Life', *Commentary*, p.56, December 1981.

Chapter 5 Who's in control?

1 Quoted in a case study prepared by Professor Jonathan West, 'E.I. duPont de Nemours and Company', Harvard Business School, 19 November 1998, No.N-9-699-037.

2 Jeremy Rifkin, *Algeny*, New York, Penguin, 1984, p.11.

3 Quoted in David Pilling, 'The Facts of Life: Chemical and Pharmaceutical Companies see their future in biological innovation', *Financial Times*, 9 December 1998, p.21.

4 W.Bratic, P.McLane and R.Sterne, 'Business Discovers the Value of Patents', *Managing Intellectual Property*, September 1998.

5 Steve Gorelick, *Small is Beautiful, Big is Subsidised*, International Society for Ecology and Culture, Dartington, Devon, October 1998, p.9.

6 'The Gene Giants: Masters of the Universe?', *RAFI Communiqué*, March/April 1999 <www.rafi.org/communique/fltxt/19992.html>.

7 As 6. Figures on the size of national economies comes from the World Bank's *World Development Report 1998/99*, World Bank, Oxford University Press, Table 1, p.190-191.

8 As 6.

9 'RAFI on Monsanto merger: Pharma-geddon', 'Geno-Types', Rural Advancement Foundation International, 21 December 1999 <www.rafi.org>.

10. 'Seedless in Seattle', RAFI News Release, Rural Advancement Foundation International, 26 November 1999.

11 Union of Concerned Scientists, 'From the Editor's Desk: Big and Bigger', *The Gene Exchange: A Public Voice on Biotechnology and Agriculture*, Summer 1998

<www.ucsusa.org/Gene/su98.big.html>.

12 US National Farmers Union, 'Action needed to halt consolidation in the agricultural industry', News Release, 26 January 1999. Cited in Rafi Communiqué in 6 above. ·

13 J.Friedland and S.Kilman, 'As Geneticists develop an appetite for greens, Mr. Romo flourishes', *Wall St. Journal*, 28 January 1999.

14 Lynn Grooms, 'With Merger Completed, Harris Moran Focuses on Future', *Seed & Crops Digest*, January 1999.

15 Personal Communication from Hope Shand, Rural Advancement Federation International, 5 January 2000.

16 R.Fraley, in *Farm Journal*. Quoted in: J.Flint, 'Agricultural industry giants moving towards genetic monopolism', Telepolis, Heise Online, 1998 <www01.ix.de/tp/english/inhalt/co/2385/1.html>.

17 Figures from Rafi Communiqué quoted in 6 above.

18 A.Coghlan, 'Gene industry fails to win hearts and minds', *New Scientist*, June 1993.

19 *Genetic Engineering: A Review of Developments in 1998*, GeneWatch Briefing Number 5, January 1999, p.4.

20 *Europe Environment* Vol.498, 22 April 1997, pp.8-9.

21 Personal communication from Sue Mayer, Director of GeneWatch, April 1999.

22 Declan Butler et al, 'Long-term effect of GM crops serves up food for thought', *Nature*, Vol.398, 22 April 1999, pp.651-656.

23 Parliamentary question from Ms. Drown to Mr. Rooker, House of Commons Hansard Written Answers, 13 May 1999 (pt 11) Question Number 83241 <www.parliament.the-stationery-office.co.uk>.

24 Stephen Gilman, 'Why Subsidize the Wrong Kinds of Plants?', Letter to the Editor, *The New York Times*, 22 April 1999.

25 Jane Merriman, 'Euro stores cash in on "Frankenstein food" fears', Reuters, London, 20 April 1999.

26 J. Toth, 'No Longer Backyard Business', *Fairfield Weekly Reader*, 607 W. Broadway, Fairfield, Iowa, USA, 11-17 September 1997.

27 H.A.Schneiderman and W.D. Carpenter, 'Planetary Patriotism: Sustainable Agriculture for the Future', *Environmental Science and Technology*, Vol.24 No.4, April 1990, p.472.

28 Quoted by Kathy Koch in the 4 September 1998 issue of *The Congressional Quarterly Researcher*.

29 Ronnie Cummins, 'S.O.S. Save Organic Standards! Round Two', *Food Bytes* No. 14 November 8, 1998 <http://www.purefood.org/Organic/foodByt14.htm>.

30 Mae-Wan Ho et al, 'The Biotechnology Bubble', *The Ecologist*, Vol.28 No.3, May/June 1998, p.146.

31 Corporate Europe Observatory, 'Smooth Façade: Greenwash Guru Burson Marsteller and the Biotech Industry', *The Ecologist*, Vol.28 No.3, May/June 1998, p.136; Sources from article available at <www.xs4all.nl/~ceo/newsletter/>; Ingeborg Woitsch, 'Manipulating consciousness with advertising strategies', translation of an article which appeared in Das Goetheanum—Wochenschrift für Anthroposophie No.30, 26 July 1998, pp.441-443 <www.anth.org/ifgene/woitsch.htm>.

32 'Communications Programmes for EuropaBio', Burson Marsteller, January 1997.

33 FACCT: A project to promote Familiarisation with and Acceptance of Crops incor-porating Transgenic Technologies in modern agriculture. A demonstration project under Framework Programme IV—European Commission. Paper OCS 8/96, Annex D & Draft Technical Annex—19 December 1995.

34 Biotechnology and Biological Science Research Council Annual Report 1996-

1997, BBSRC, Swindon.
35 'Genetic Engineering: Can it Feed the World?', GeneWatch Briefing No.3, p.1.
36 'BBSRC Scientists Gagged on GM Foods', Norfolk Genetic Information Network Press Release, 19 February 1999 <http://members.tripod.com/~ngin/>.
37 Henry Thomas Stelfox et al, 'Conflict of Interest in the Debate over Calcium-Channel Antagonists', New England Jnl of Medicine, Vol.338, No.2, 8 January 1998, pp.101-106.
38 Richard A. Knox, 'Study finds conflict in medical reports', Boston Globe, 8 Jan 1998.
39 Peter Montague, 'Follow The Money', Rachel's Environment & Health Weekly, No.581, 15 January 1998 <www.monitor.net/rachel/r581.html>.
40 Environment News Service, London, 20 June 1997
<www.envirolink.org/environews/ens/>.
41 Kathleen Hart, 'House Agriculture subcommittees question EPA's authority to regulate biotech plants as pesticides', Pesticide Toxic Chemical News, 25 March 1999.
42 'What is the FDA Policy for Regulation of Genetically Engineered Foods?', The Council for Responsible Genetics <www.essential.org/crg/consumeralert.html>; Susan Wright, 'Splicing Away Regulations—Down on the Animal Pharm' <www.purefood.org/pharm.html>.
43 Figures from Nature Biotechnology, August 1997, cited in Stephen Nottingham, Eat Your Genes, Zed Books, London, 1998, p.172.
44 Speech by the Rt. Hon. Stephen Byers MP, Secretary of State for Trade and Industry, to Biotechnology Industry Association Gala Dinner, 21 January 1999 <www.dti.gov.uk/public/search.html>.
45 René von Schomberg, 'An appraisal of the working in practice of directive 90/220/EEC on the deliberate release of genetically modified organisms', Scientific and Technological Options Assessment (STOA) of the European Parliament, 2 January 1998 <www.europarl.eu.int/dg4/stoa/en>.
46 The Guardian, 7 December 1996, p.10. Cited in Stephen Nottingham, Eat Your Genes, Zed Books, London, 1998, p.138.
47 Geoffrey Lean, 'Third World rejects GM Environment' Independent on Sunday, London, 28 February 1999.
48 Personal communication with Dr. Tewolde Berhan Gebre Egziabher, Institute for Sustainable Development, Box 30231, Addis Ababa, Ethiopia, 7 February 1999.
49 'International Committee Rejects Consumer Call for Mandatory Labelling of Genetically Engineered Food', Consumers International Press Release, 28 May 1998 <http://193.128.6.150/consumers//news/pressreleases/codex270598.html>.
50 Ben Lilliston and Ronnie Cummins, 'Organic Vs "Organic": The Corruption of a Label', The Ecologist, Vol.28, No.4, July/August, 1998, p.197.
51 'US warns EU not to impede farm trade over biotech', Reuters, Washington, 19 June 1997.
52 Environment News Service, London, 20 June 1997
<www.envirolink.org/environews/ens/>.
53 Leila Corcoran, Reuters, Washington, 17 July 1997
<www.geocities.com/Athens/1527/egypt.html>.
54 'Japan may require labels on genetic food', Nature, Vol.395, No.628, 15 October 1998; 'No Support For Genetic Engineering', Positive News, No.19, Spring 1999.
55 Cabinet Minutes from the New Zealand Government, 19 February 1998. Reported in a feature article in UK newspaper the Independent on Sunday, 22 Nov 1998.
56 Ronnie Cummins & Ben Lilliston, Campaign for Food Safety News No. 22, 21 October 1999 <www.purefood.org>.
57 Union of Concerned Scientists, 'What's Coming to Market?', The Gene Exchange:

A Public Voice on Biotechnology and Agriculture, Fall/Winter 1998
<www.ucsusa.org/Gene/w98.market.html>.
58 *Genetic Engineering: A Review of Developments in 1998,* GeneWatch Briefing
Number 5, January 1999, p.2.
59 'GM could be used in up to 90% of processed food, says GeneWatch',
GeneWatch Press Release, 15 February 1999; Information on 'Available enzymes
made by genetically modified microorganisms for use in food processing' sourced
from the Association of Manufacturers of Fermentation Enzyme Products.
60 Quoted in *Splice,* The Magazine of the Genetics Forum, Vol.4, Issue 6, Aug/Sept 1998.
61 Quoted by Kathy Koch in the September 4th, 1998 issue of the *Congressional
Quarterly Researcher.*
62 D.Evans, 'Produce-on-demand: What's good for US markets is good for world
markets too', *Nature Biotechnology* Vol.14, 1996, p.802.
63 Marvin Hayenga, 'Structural Change in the Biotech Seed and Chemical Industrial
Complex', *AgBioForum,* Vol.1 No.2, Fall 1998; 'Genetic Engineering: A Review of
Developments in 1998', *GeneWatch Briefing Number 5,* January 1999, p.6; also as 6.
64 J. West, 'E.I. du Pont de Nemours and Company', Harvard Business School Case
Study, N-9-699-037, 19 November 1998, p.8.
65 V. Brower, 'Nutraceuticals: poised for a healthy slice of the healthcare market?',
Nature Biotechnology, Vol 16, pp.728-731.
66 'The Facts of Life', *Financial Times,* 9 December 1998.

Chapter 6   A Case Study: Milk and GE Growth Hormones

1 Status Update from Monsanto POSILAC(r) bovine somatotropin December 15,
1998 <www.monsanto.com/dairy/press4.html>
2 *Extra!* Update, June 1998 <www.fair.org/extra/9806/foxbgh.html>.
3 J. Batz, 'Hormonal Rage: Monsanto spikes a Florida TV story about its bovine growth
hormone' <www.geocities.com/Athens/1527/rBGH/foxBgh.htm>.
4 Sheldon Rampton and John Stauber, 'This Report Brought to you by Monsanto',
The Progressive, July 1998, pp.23-24.
5 'Reporters Blow Whistle On News Station' <www.foxBGHsuit.com/index2.htm>.
6 As 5; also as 3.
7 P. Montague, 'Milk, rBGH, and Cancer', *Rachel's Environment & Health Weekly*
No.593, 9 April 1998 <www.geocities.com/Athens/1527/rBGH/rach593.html>.
8 Sheldon Rampton and John Stauber, 'This Report Brought to you by Monsanto',
The Progressive, July 1998, pp.23-24.
9 'Reporters Blow Whistle On News Station' <www.foxBGHsuit.com/index2.htm>.
10 Press release by Steve Wilson and Jane Akre, 10 September 1998
<www.foxBGHsuit.com/jasw0910.htm>.
11 M. Hansen, 'Consumers Union/Consumer Policy Institute statement on Canadian
decision to ban use of recombinant bovine growth hormone on dairy cows',
Consumer Policy Institute, Yonkers, NY, 15 January 1999.
12 K. Schneider, 'FDA Warns the Dairy Industry Not to Label Milk Hormone-Free',
New York Times, 8 February 1994.
M. Taylor, 'Interim Guidance on the Voluntary Labelling of Milk and Milk Products
From Cows That Have Not Been Treated With Recombinant Bovine Somatotropin',
Federal Register Vol.59 No.28, 10 February 1994, pp.6279-6280.
13 P. Montague, *Rachel's Hazardous Waste News,* No.381, 17 March 1994

<www.enviroweb.org/pubs/rachel/rhwn381.htm>.

14 'Two New Deputy Commissioners Named By Kessler', FDA Talk Paper [T91-38], 15 July 1991.

15 As 13; also 'The rBGH Scandals' <www.geocities.com/Athens/1527/text4.html>; Stephen P. Rosenfield, 'Monsanto, dairies battle over free speech', The Times-Argus, Montpelier, Vermont, 22 February 1994.

16 W. Hobgood, Vice President, Monsanto Company Animal Sciences Division, Letter to Retail Grocer, 18 February 1994; Robert Steyer, 'BST has the Mail moving on ads: Monsanto writes warning letters', St. Louis Post Despatch, 24 February 1994.

17 'Monsanto's Dirty Tricks', Grassroots, The Pure Food Campaign, October 1995, p.13; 'Statement on Misleading Promotion and Advertising Activities', an anonymous statement on Monsanto letterhead dated 4 March 1994, faxed to Rachel's Hazardous Waste News by staff of Tom McDermott of Monsanto in St. Louis, MO, 'Some Dangers of Hormones in Milk', Rachel's Hazardous Waste News, No.382, 24 March 1994 <www.enviroweb.org/pubs/rachel/rhwn382.htm>; 'Carrying Monsanto's Water', Rural Vermont Report, Summer 1998.

18 R. Collier et al, Letter to the Editor, The Lancet, 17 Sept 1994, Vol.344, p.816.

19 Cited in Gail Feenstra's Introduction to William C. Liebhardt, The Dairy Debate; Consequences of Bovine Growth Hormone and Rotational Grazing Technologies, University of California Sustainable Agriculture Research and Education Program, Davis, CA, 1993, pp.20-23.

20 D. Challacombe and E. Wheeler, 'Safety of Milk from Cows Treated With Bovine Somatotropin', The Lancet, 17 September 1994, vol.344, p.815.

21 T. Mepham et al, 'Safety of milk from cows treated with Bovine Somatotropin', The Lancet, Vol.334, 19 November 1994, pp.1445-1446.

22 C. Xian, 'Degradation of IGF-1 in the Adult Rat Gastrointestinal Tract is Limited by a Specific Antiserum or the Dietary Protein Casein', Journal of Endocrinology, Vol.146, No.2, 1 August 1995, pp.215-225; R. Rao et al, 'Luminal Stability of Insulin-Like Growth Factors I and II in Developing Rat Gastrointestinal Tract', Journal of Pediatric Gastroenterology and Nutrition, Vol.26, No.2, February 1998, pp.179-185; T. Kimura et al, 'Gastrointestinal Absorption of Recombinant Human Insulin-Like Growth Factor-I in Rats', Journal of Pharmacology and Experimental Therapeutics, Vol.283, No.2, November 1997, pp.611-618; D. Burrin et al, 'Orally administered IGF-I increases intestinal mucosal growth in formula-fed neonatal pigs', American Journal of Physiology, Vol.270, No.5 Part 2, May 1996, pp.R1085-R1091; R. Philipps, 'Growth of artificially fed infant rats: effect of supplementation with insulin-like growth factor I', American Journal of Physiology, Vol.272, No.5 Part 2, May 1997, pp.R1532-R1539.

23 J. Holly, 'Insulin-like growth factor-I and new opportunities for cancer prevention,' Lancet Vol.351, No.9113, 9 May 1998, pp.1373-1375.

24 S. Hankinson et al. 'Circulating concentrations of insulin-like growth factor I and risk of breast cancer', Lancet Vol.351, No.9113, 9 May 1998, pp.1393-1396; J. Holly, 'Insulin-like growth factor-I and new opportunities for cancer prevention', Lancet, Vol.351, No.9113, 9 May 1998, pp.1373-1375; S. Epstein, 'Comments from the Cancer Prevention Coalition on IGF-1 and cancer based on articles from the January 23rd 1998 edition of Science magazine and the May 9th edition of The Lancet', The Ecologist, Vol.28, No.5, Sept/Oct 1998, pp.268-269; Cancer Prevention Coalition, 'Researcher Warns of Cancer Risk From rBGH (non-organic) Dairy Foods' <www.holisticmed.com/bgh/prostate.html>.

25 'Growth Hormones: A Case of Regulatory Abdication', International Journal of

Health Services, Vol.26, No.1, 1996, pp.173-185; S. Epstein, 'Comments from the Cancer Prevention Coalition on IGF-1 and cancer based on articles from the January 23rd 1998 edition of Science magazine and the May 9th edition of The Lancet', *The Ecologist*, Vol.28, No.5, Sept/Oct 1998, pp.268-269; J. Chan et al, 'Plasma Insulin-Like Growth Factor-I and Prostate Cancer Risk: A Prospective Study', *Science*, Vol.279, 23 January 1998, pp.563-566.

26 Clip from Monsanto's sales tape for Posilac (rBST) 'The Mystery in Your Milk', Reporters' Version Part II <www.foxBGHsuit.com/index2.htm>

27 Frederick Bever, 'Canadian Agency Questions Approval of Cow Drug by US', Associated Press, *Rutland Herald* (Vermont), 6 October 1998.

28 Samuel S. Epstein, 'Unlabelled milk from cows treated with biosynthetic growth hormones: a case of regulatory abdication', *International Journal of Health Services*, Vol.26, No.1, 1996, pp.173-185.

29 M. Greger, 'What You May Not Know' <http://arrs.envirolink.org/AnimaLife/spring95/BGH.html>.

30 'The Tainted Milk Mustache—How Monsanto and the FDA Spoiled a Staple Food', *Alternative Medicine Digest*, Jan 1999 <www.afpafitness.com/MilkMustache.htm>; Michael Greger, 'Bovine Growth Hormone—What You May Not Know' <http://arrs.envirolink.org/AnimaLife/spring95/BGH.html>.

31 M. Hansen, 'Consumers Union/Consumer Policy Institute statement on Canadian decision to ban use of recombinant bovine growth hormone on dairy cows', Consumer Policy Institute, Yonkers, NY, 15 January 1999.

32 J. Juskevich and C. Guyer, 'Bovine Growth Hormone: Human Food Safety Evaluation', *Science*, Vol.249, 1990, pp.875-884.

33 As 31.

34 S. Chopra et al, 'rBST (Nutrilac) Gaps Analysis Report by rBST (Nutrilac) Internal Review Team', Health Protection Branch, Health Canada, Ottawa, 21 April 1999 <www.nfu.ca/nfu/Gapsreport.html>.

35 As 32, p.878.

36 F. Bever, 'Canadian Agency Questions Approval of Cow Drug by US', Associated Press, *Rutland Herald* (Vermont), 6 October 1998.

37 As 32, p.876.

38 'The rBGH Scandals' <www.geocities.com/Athens/1527/text4.html>.

39 A. McIlroy, 'Health Canada cover-up alleged', *Toronto Globe and Mail*, 17 September 1998.

40 L. Eggertson, 'Researchers threatened, inquiry told ', *Toronto Star*, 17 Sept 1998.

41 P. Montague, 'Milk, rBGH, and Cancer', *Rachel's Environment & Health Weekly*, No.593, 9 April 1998 <www.geocities.com/Athens/1527/rBGH/rach593.html>; 'The rBGH Scandals' <www.geocities.com/Athens/1527/text4.html>.

42 'The Mystery in Your Milk', Reporters' Version Part I, Version 29 <www.foxBGHsuit.com/index2.htm>; A. McIlroy, 'Ottawa Refuses to Approve Bovine Growth Hormone', *Globe and Mail* (Canada), 20 February 1999; 'Canada to reject Monsanto growth hormone—report', Reuters, Ottawa, 14 January 1999.

43 Laura Eggertson, 'Expert worked for drug firm', *Toronto Star*, 21 September 1998.

44 See full report at <www.hc-sc.gc.ca/english/archive/rbstanimals/02.htm>; M. Hansen, 'Consumers Union / Consumer Policy Institute statement on Canadian decision to ban use of recombinant bovine growth hormone on dairy cows', Consumer Policy Institute, Yonkers, NY, 15 January 1999; also as 34.

45 'Report on Animal Welfare Aspects of the Use of Bovine Somatotrophin Report of the Scientific Committee on Animal Health and Animal Welfare Adopted 10 March

1999'. The Working Group included Prof. D. Broom, Dr. R. Dantzer, Prof. P. Willeberg, Prof. B. Mepham and Prof. E Noordhuizen-Stassen
<http://europa.eu.int/comm/dg24/health/sc/scah/out21_en.html>.

46 William M. Murphy and John R. Kunkel, 'Sustainable Agriculture: Controlled Grazing vs. Confinement Feeding of Dairy Cows', in William C. Liebhardt, 'The Dairy Debate; Consequences of Bovine Growth Hormone and Rotational Grazing Technologies', University of California Sustainable Agriculture Research and Education Program, Davis, CA, 1993, p.121; also as 31 and 34.

47 As 46.

48 As 32, p.875.

49 As 32, p.875.

50 See full report at <www.hc-sc.gc.ca/english/archive/rbstanimals/02.htm>; also as 31 and 34

51 As 34.

52 E. Millstone, E. Brunner and I. White, 'Plagiarism or Protecting Public Health?' Nature, Vol.391 No.6499, 20 October 1994, pp.647-648; D. Kronfeld, 'Safety of Bovine Growth Hormone', Science, Vol.251, 18 January 1991, pp.256-257; T. Mepham, 'Public health implications of bovine somatotropin use in dairying: discussion paper', Journal of the Royal Society of Medicine, Vol.85, December 1992, pp.736-739; E. Chelimsky et al, 'Recombinant Bovine Growth Hormone: FDA Approval Should Be Withheld Until the Mastitis Issue is Resolved' [GAO/PEMD-92-26], US General Accounting Office, Gaithersburg, MD, 1992; P. Montague, 'Some Dangers of Hormones in Milk', Rachel's Hazardous Waste News, No.382, 24 March 1994 <www.enviroweb.org/pubs/rachel/rhwn382.htm>.

53 James Ridgeway, 'Robocow—How Tomorrow's Farming is Poisoning Today's Milk', The Village Voice, 14 March 1995.

54 As 34.

55 M. Greger, 'What You May Not Know'
<http://arrs.envirolink.org/AnimaLife/spring95/BGH.html>.

56 M. Hansen, 'Biotechnology and Milk; Benefit or Threat? An Analysis of Issues Related to BGH/BST Use in the Dairy Industry', Consumer Policy Institute/ Consumers Union, Mount Vernon, NY, 1990, p.1; P. Montague, Rachel's Hazardous Waste News, No.384, 7 April 1994.

57. 'U.S. and Europe Agree to Disagree on Safety of Dairy Hormone,' Consumer Policy Institute Press Release, June 30, 1999
<http://www.consumer.org/food/bghny899.htm>; 'Codex Alimentarius delays decision on BST/rBGH', FAO Clarification note,
<www.fao.org/es/ESN/codex/BST_Fact.htm>

58 Toxicological Evaluation of Certain Veterinary Drugs Residues in Food, World Health Organisation Food Additive Series No.31, WHO, Geneva, 1993, p.149.

59 G. Palast, 'Outrage over Monsanto's underhand tactics in EU', The Observer, 14 March 1999.

60 'Bovine somatotropin—who's crying over spilt milk?', The Lancet, 23 January 1999.

61 N. Avery, M. Drake and T. Lang, 'Cracking the Codex: An Analysis of Who Sets World Food Standards', The National Food Alliance, London, 1993; 'Rejection of Hormones for Milk Production Applauded by International Consumer Group', Consumers International Press Release on rBGH (rBST), 3 July 1997,
<www.geocities.com/Athens/1527/rejrbgh.html>.

62 Alan Simpson MP, House of Commons Hansard Debates, 22 March 1999 (pt 29) <www.parliament.the-stationery-office.co.uk/cgi-bin/tso_fx>.

Chapter 7 **Turning the Tide**

1 Greenpeace Press Release,'Leaked Document from Monsanto Reveals Collapse of Public Support For Genetically Engineered Foods', 18 November 1998
2 Based upon a $9 billion annual revenue for Monsanto (David Usborne, 'The Monster within Byline', *The Independent*, London, 21 April 1999) and a $31.916 million annual income for Greenpeace International (Greenpeace International pro forma summary financial statements, year ending 31 December 1997). The annual revenue for Greenpeace International is 0.35% of the annual revenue of Monsanto.
3 Genetix Update, March 99.
4 Susie Steiner, 'Protesters are cleared over gene crop raid', *The Times*, 30 March 1999.
5 'Campaigners declare victory as government set to back out of genetics trial', Genetic Engineering Network Media Release, 29 March 1999.
6 'Monsanto is costing me £6000 a year!', *The Food Magazine*, November 1997, p.3.
7 Personal communication from Anne Ward, April 1999.
8 Zac Goldsmith, 'The Monsanto Test', *The Ecologist*, Vol.29 No.1, Jan/Feb 1999, p.5.
9 Geoffrey Lean, 'GM foods—Victory for grass-roots action', *Independent on Sunday*, London, 2 May 1999.
10 Taken from a series of quotes on the *Natural law Party Wessex Web page*, GE section <www.btinternet.com/~nlpwessex/Documents/contentsfall.htm>.
11 Isabelle Meister and Thomas Schweiger, 'Europe Questions GMO Releases: Chronology of growing resistance to GMOs', Greenpeace International, April 1999.
12 Mauro Zanatta and Katia Guimaraes, 'Agriculture specialists request suspension of transgenic seed production' Gazeta Mercantil, Brazil, 16 April 1999.
13 As 11.
14 'Government Association Supports GM Food Ban', *Friends of the Earth Biotech Mailout*, Vol 5, Issue 2, 15 March 1999
<www.foeeurope.org/programmes/biotechnology/5n2_frames.htm>.
15 *Independent on Sunday*, London, 27 January 1999; 'Top restaurants back genetic food ban,' *Friends of the Earth Press Release*, 30 December 1998; 'Guess What You've Been Eating', *Sydney Morning Herald*, Australia, 12 December 1998.
16 Alex Kirby, 'Monsanto's caterers ban GM foods', BBC Online, 22 December 1999, <http://news.bbc.co.uk/hi/english/sci/tech/newsid_574000/574245.stm>.
17 As 11.
18 'Britain's biggest farmer pulls out of gene trials', Reuters, 30 March 1999; Jonathan Petre, 'Church bans GM crop trials on its land', *Daily Telegraph*, 5 December 1999.
19 Simon de Bruxelles, 'I killed GM crop with a heavy heart', *The Times*, 8 June 1999.
20 'GM protesters arrested as crops ripped up', BBC News Online, Sunday, 18 July 1999 <http://news.bbc.co.uk/hi/english/uk/newsid_397000/397865.stm>.
21 Karnataka State Farmers Association Press Release, 'Monsanto's Cremation Starts in Karnataka', Sindhanoor, India, 28 November 1998
22 Personal Communication from Jim Thomas, Greenpeace UK, 29 December 1999
23 *Genetic Engineering: A Review of Developments in 1998*, GeneWatch Briefing No.5, Jan 1999, p.3; also as 11.
24 'Greenpeace welcomes GMO moratorium in EU but condemns double standards for developing countries', Greenpeace International Press Release, 25 June 1999.
25 Alex Kirby, 'US court challenge to Monsanto', BBC News Online, 14 December 1999 <http://news2.thls.bbc.co.uk/hi/english/sci/tech/newsid_564000/5649830.stm>.
26 Greenpeace Press Release,'Leaked Document from Monsanto Reveals Collapse of Public Support For Genetically Engineered Foods', 18 November 1998.
27 Quoted in David Korten, *The Post-Corporate World*, Kumarian Press, 1999, p.225.

Index

CHELSEA GREEN

Sustainable living has many facets. Chelsea Green's celebration of the sustainable arts has led us to publish trend-setting books about organic gardening, solar electricity and renewable energy, innovative building techniques, regenerative forestry, local and bioregional democracy, and whole foods. The company's published works, while intensely practical, are also entertaining and inspirational, demonstrating that an ecological approach to life is consistent with producing beautiful, eloquent, and useful books, videos, and audio cassettes.

For more information about Chelsea Green, or to request a free catalog, call toll-free (800) 639-4099, or write to us at P.O. Box 428, White River Junction, Vermont 05001. Visit our Web site at www.chelseagreen.com.

Chelsea Green's titles include:

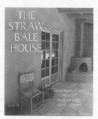

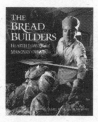

The Straw Bale House
The Independent Home
The Rammed Earth House
The Passive Solar House
The Sauna
Wind Energy Basics
The Solar Living Sourcebook
A Shelter Sketchbook
Mortgage-Free!
Hammer. Nail. Wood.

The Bread Builders: Hearth Loaves and Masonry Ovens
The Apple Grower
The Flower Farmer
Passport to Gardening: A Sourcebook for the 21st-Century
The New Organic Grower
Four-Season Harvest
Solar Gardening
Straight-Ahead Organic
The Contrary Farmer

Whole Foods Companion
Gaviotas: A Village to Reinvent the World
Who Owns the Sun?
Global Spin: The Corporate Assault on Environmentalism
Hemp Horizons
Beyond the Limits
The Man Who Planted Trees
Simple Food for the Good Life